Decision Making in Timber Production, Harvest, and Marketing

Clawson explores the issues related to timber management with a particular focus on the harvesting of timber stands in *Decision Making in Timber Production, Harvest and Marketing*. Originally published in 1977, her study considers biological, economic and management implications of timber growing as well as the decision-making process in U.S forest situations including methods of analysis. This title will be of interest to students of Environmental Studies and professionals.

Decision Making in Timber Production, Harvest and Marketing

Marion Clawson

RFF PRESS
RESOURCES FOR THE FUTURE

First published in 1977
by Resources for the Future, Inc.

This edition first published in 2016 by Routledge
2 Park Square, Milton Park, Abingdon, Oxon, OX14 4RN
and by Routledge
711 Third Avenue, New York, NY 10017

Routledge is an imprint of the Taylor & Francis Group, an informa business

Publisher's Note
The publisher has gone to great lengths to ensure the quality of this reprint but points out that some imperfections in the original copies may be apparent.

Disclaimer
The publisher has made every effort to trace copyright holders and welcomes correspondence from those they have been unable to contact.

A Library of Congress record exists under LC control number: 77084930

ISBN 13: 978-1-138-94952-2 (hbk)
ISBN 13: 978-1-315-66898-7 (ebk)

Decision making in timber production, harvest, and marketing

MARION CLAWSON

RESEARCH PAPER R-4

RESOURCES FOR THE FUTURE / WASHINGTON, D.C.

RESOURCES FOR THE FUTURE, INC.
1755 Massachusetts Avenue, N.W., Washington, D.C. 20036

Resources for the Future is a nonprofit organization for research and education in the development, conservation, and use of natural resources and the improvement of the quality of the environment. It was established in 1952 with the cooperation of the Ford Foundation. Grants for research are accepted from government and private sources only if they meet the conditions of a policy established by the Board of Directors of Resources for the Future. The policy states that RFF shall be solely responsible for the conduct of the research and free to make the research results available to the public. Part of the work of Resources for the Future is carried out by its resident staff; part is supported by grants to universities and other nonprofit organizations. Unless otherwise stated, interpretations and conclusions in RFF publications are those of the authors; the organization takes responsibility for the selection of significant subjects for study, the competence of the researchers, and their freedom of inquiry. This book is a product of RFF's Energy and Materials Division, directed by Hans H. Landsberg and Sam H. Schurr.

Library of Congress Catalog Card Number 77-84930
ISBN 0-8018-2048-0

Manufactured in the United States of America

Published September 1977. $5.75.

PREFACE

This report considers timber management, especially the harvest of timber stands, in a broad context. The biological, economic, and management considerations of timber growing are all considered; each is important but none is adequate alone to serve as a basis for the actual management of forests. The important interrelations among timber, recreation, wilderness, wildlife, water, and other outputs of forests are not considered in this report, in part because I have analyzed them elsewhere. This report is concerned with U.S. forest situations; those in some other countries may be different.

This report was conceived and has been written in the conviction that foresters, economists, conservationists, and business interests were not talking in the same terms when they considered the growing and the harvest of timber from particular forest areas. I hope to clarify some of the problems, relationships, and opportunities of timber growing by a more comprehensive and eclectic approach than has been taken hitherto.

This report considers the decision-making process in timber management and harvest for the small private forest, for the large integrated forest industry firm, and for the national forests. For the latter, the special problems of managing inherited old growth timber stands, of sustained yield, of even flow, and of community stability are each considered. The opportunities and problems of national forest timber management under the National Forest Management Act of 1976, are given particular attention although any analysis prior to several years operations under that Act is necessarily somewhat speculative. The focus throughout this report is on methods of analysis and on kinds of considerations which should enter into decision-making; no attempt is made, and none would be possible in a relatively brief report, to analyze in detail a number of specific situations or cases.

This report has been written for a broad audience. It is addressed to foresters, economists, conservationists, businessmen, legislators, and the general public. Hopefully, each will find something of use and value for his or her special interests; and hopefully each will gain a better understanding of how certain problems may be perceived by other groups with different backgrounds and different interests. Hopefully, each will gain a better understanding of the particular problems, opportunities, and limitations of timber-growing by small and large private forest owners and on the national forests.

The style and the level of analysis have been chosen with the intended audience in mind. That is to say, the report is not highly technical in its terminology, its form of analysis, or even in its concepts. Considerations of biology and of economics are introduced but in terms which can readily be understood by the average concerned person. The more technical matter is included in an appendix. I hope and believe that the analysis is significant but the methods of presentation are not complex.

Although this report was not written explicitly for college students majoring in forestry, I hope they too will find much of interest and value here. I believe the approaches used here may be useful supplements to customary courses in silviculture, timber management, economics, and other somewhat specialized courses. My concern has been with practical problems of decision-making by the forest owner or manager, and my viewpoint is that a considerable number and range of considerations are likely to enter into such decisions.

In writing this report, I have been influenced by many persons, including, I am sure, some whose influence I am unaware. Where I have drawn specifically on the writings of others, this is acknowledged directly in the text. I am particularly indebted to several of my colleagues at Resources for the Future, notably Emery N. Castle and William F. Hyde. The latter made available to me an extensive manuscript, unpublished at the date of this writing. Reference is made to it in the text at appropriate places. It has been most helpful to me. An earlier draft of this report was read and commented upon by Frank J. Convery, Duke University;* Barney Dowdle, University of Washington;* Mason Gaffney, University of California, Riverside;* Hans M. Gregersen, University of Minnesota;* Charles W. Howe, University of Colorado;* Lawrence W. Libby, Michigan State University;* John H. Muench, National Forest Products Association;* Dennis Teeguarden, University of California, Berkeley;* and Thomas R. Waggener, University of Washington*. I revised the earlier draft in light of their comments. Not all comments by all reviewers were incorporated into the present draft, in part because some comments contradicted others and in part because I did not agree with some. The assistance of these reviewers has been very great and I am happy to take this occasion to express my thanks to them. None of the persons mentioned here, of course, is responsible for my final statement of the material contained herein.

Marion Clawson
Washington, D.C.
March 1977

*Organization given for identification only; each responded as an individual.

Contents

List of Figures

List of Tables

DECISION-MAKING IN TIMBER PRODUCTION, HARVEST, AND MARKETING

Forests produce many goods and services of use to man--water, wildlife, recreation, and wilderness, as well as wood or timber which is often the output which produces all or nearly all of the cash income from the forest operation. Decisions upon the proper mix of forest outputs, and upon the relation of specific forest site and other characteristics to those decisions, are important aspects of forestry in its broadest terms, as I have pointed out elsewhere [1][2]. This report is more limited in its scope. It is concerned with the production of timber or wood only--how to operate the forest for economic wood production, including particularly such matters as the time to harvest, the best way to market output, and the relations of forest production to the processing and marketing of the wood products. Concentration upon this particular range of issues does not deny the importance of the other forest outputs, but for any forest area which is to grow trees for harvest, the kinds of considerations discussed in this report are important.

Characteristics of Timber Growing

The growing of timber as an economic or business enterprise has many similarities and some differences from most other natural resource uses. Forestry involves physical and biological relationships, as do other natural resource uses. Soils, climate, and other natural factors affect tree growth; input-output relationships, or production functions, which relate wood output to the inputs of capital, land, labor, and management are critical here as in most other natural resource uses. There are complementary and competitive relationships among the species of trees, between trees and other vegetation, and between trees and insects and diseases. There is an energy flow and an energy balance, likewise a moisture flow and balance, and forms of nutrient cycles. In all this, tree growing has many similarities to the growing of cultivated farm crops. Also, there are environmental effects, some quite favorable, others less so, from the growing and the harvest of trees.

Timber growing has also considerations of economic efficiency similar to those for many other natural resource uses--how great are the costs, how large are the returns, what is the balance between costs and returns, what is the optimum scale and the most desirable method of production, and the like. There are also questions of economic equity involved in timber growing, as in much other natural resource use--who pays the costs, and who receives the gains? [1]. Also like many other resource uses, timber growing frequently has a marked degree of disassociation of cost and return--one group pays much of the cost, another group gets much of the gain. There are also cultural or social aspects to timber growing--some methods offend people, or at least some people, regardless of their biological or economic efficiency. And, lastly, timber growing involves matters of administrative or operational practicality--can the methods which seem most efficient or most desirable actually be carried out in practice? All of this has been dealt with elsewhere, and need not be explored in detail in this report, and with some modest differences all of this is not unique to forestry but is common for many natural resource situations.

But the production of wood and timber in forests has some attributes which are unique in kind or in degree or at least are not widely shared among other natural resource uses.

First, the same physical entity--the tree--is both capital and output. While the tree is allowed to grow, so that annually it produces more wood, then it is productive capital, in many respects similar to a machine in a factory. When the tree is harvested, its wood is output. Trees normally appreciate in value during their growing years, but decreases in value due to physical destruction (fire, windthrow, ice storms, insect and disease infestation, etc.) are by no means uncommon. Forestry would be a lot simpler and more profitable if value-appreciation during growth was invariably encountered. Typically, trees are harvested when some years of age so that the production of a considerable number of past years becomes marketable output all at once. At any given date, the forest owner or manager must decide whether to let the tree grow another year or longer, or whether to harvest it now. Decisions about when to harvest are some of the most important management decisions in forestry. If the owner-manager decides to harvest, he receives some income, which he can reinvest in the forest or can invest elsewhere or can consume; if he decides not to harvest, he foregoes this income until a later date when he does harvest, and interest on the foregone capital is one of the major costs of growing more wood. We shall return to this point in much more detail later in this report.

In this characteristic of the same physical substance being both capital and output, timber growing resembles cattle growing. The heifer or cow retained in the breeding herd is capital, kept to produce a calf or calves, while the same animal when sold is output. The cycle of cattle production is shorter, since a cow is rarely productive beyond perhaps twelve years of age, while a timber stand may grow, though slowly, for well over one hundred years. These are often called appreciating assets because value typically increases with age, up to some point. But the value of any of them may decline due to one or more of several factors.

Second, the wood grown in the forest is a raw material, not a consumption good. The standing tree is a raw material comparatively far from consumer use; when cut, it becomes a log which must be transported to a processing plant; there it becomes lumber, plywood, pulp (for paper), or some other raw material product. These raw materials are used in making consumption goods and services. In this respect, forests and timber growing have some similarities with the petroleum industry--there are sources of supply, there must be a processing stage, and there is need for a marketing system to get the raw material from its source to its users.

Forestry is also notable for its long production period, with the great uncertainty this involves as to future costs and returns, for the unavoidable off-site environmental effects associated with even the best timber harvest, and for its very extensive use of land.

The significance of these similarities and differences with other natural resource uses may become clearer as the analysis of this report progresses.

Time-Growth-Volume Relationships for Timber Production

A tree is a biological organism and as such exhibits typical birth, growth, maturity, decay, and death processes common to all biological organisms. The tree starts very small, as a seed in natural forest reproduction. If the seed germinates, a seedling is established, and if this survives, it goes successively through sapling, pole, early adult and mature adult stages. The natural reseeding of a forest area after some severe disturbance such as fire or severe storm, or after man-organized harvest, is often highly variable in terms of numbers of seedlings, from none to several thousand per acre, and in time from a few months to several years. In well managed forests, seedlings are grown in nurseries (indoors or out-of-doors) and planting stock, typically two years of age, is produced and planted. The two year old seedling is approximately the size of a pencil for most species. If planted on a 5' X 5' spacing, or 25 square feet per seedling, then the 43,560 square feet in the acre provide space for 1,742 seedlings; if on a 7' X 7' spacing (a fairly common commercial practice) 889 seedlings would be planted per acre; and if on a 10' X 10' spacing, 436 per acre. By the time the tree had reached adulthood, it is likely that there will be about 100 stems per acre remaining--the others having died from one cause or another during the growth process.

A tree is an immobile organism, fixed in location; it cannot shift its location to a more favorable spot, but must endure the soil, moisture, fire, storm, disease, and insects which time sends to it. The marvel is that any trees survive these hazards. Competition among trees for light, moisture, and fertility eliminates many. As noted, even under favorable conditions only a comparatively few of the trees in the original stand survive to adulthood. The mature trees ultimately dies, the victim of one of the numerous hazards to tree growth, or simply from old age. At every stage in a tree's growth from young adulthood onward, an outer ring of wood on the tree trunk or stem, generally varying from one to three inches thick, is sapwood, alive and performing a vital function in transport of moisture and plant foods between roots and leaves; and within this lies the heart wood, dead but still useful to the tree. The heart wood may be highly valuable for man's use, or it may begin to rot and decay before the tree is harvested. After the tree dies, it may continue to stand, often rotting very slowly and sometimes still harvestable and usable by man, or it may fall after which rotting takes place much more rapidly--but still, for some species, to be measured in decades rather than in years.

These general relationships are put in quantitative terms for Douglas fir in one defined site situation in table 1. It must be emphasized that these figures are physical volumes of wood, not values of wood; that is, the curve of total value per acre might diverge somewhat from the curve of total physical volume per acre, if wood of some ages was more valuable than wood of other ages. The data in the table refer to a "fully stocked" stand of one acre in extent--that is, they do not relate to a single tree. Natural reproduction frequently results in sparse stands, or in uneven distrubution of seedlings over the surface of the land so that growth of some is hampered by crowding while on other areas some land surface is too poorly covered by trees to take advantage of the sun, moisture, and fertility of the site. Full stocking might occur naturally or it might result from planting seedlings. The data in the table provide information on surviving stand at each age. This is net growth to that age, the trees which have died not being counted. The figures do not tell us how many trees per acre have survived at any age. The data start with a date when

Table 1. Time-growth-volume relationships for Douglas fir, site class II, site index 180[a]

age (years)	Volume standing trees over 5.0 inches d.b.h. 1/ cu.ft.	Mean annual net growth to age 2/		Change in mean annual net growth over decade cu.ft. 3/	Increase in standing volume over preceding decade		
		Volume cu.ft.	as % of volume at age		total 4/ cu.ft.	average annual volume cu.ft. 5/	average annual as % of volume at end of decade 5/
	(1)	(2)	(3)	(4)	(5)	(6)	(7)
20	1,190	59.5	5.0	-	-	-	-
30	3,600	120.0	3.3	+60.5	2,410	241	6.69
40	6,400	160.0	2.5	+40.0	2,800	280	4.38
50	8,850	177.0	2.0	+17.0	2,450	245	2.77
60	11,050	184.2	1.7	+ 7.2	2,200	220	1.99
70	12,850	183.6	1.4	- 0.6	1,800	180	1.40
80	14,500	181.2	1.2	- 2.4	1,650	165	1.14
90	15,900	176.7	1.1	- 4.5	1,400	140	0.88
100	17,100	171.0	1.0	- 5.7	1,200	120	0.70
110	18,000	163.6	.9	- 7.4	900	90	0.50
120	18,800	156.7	.8	- 6.9	800	80	0.43
130	19,450	149.6	.77	- 7.1	650	65	0.33
140	20,000	143.6	.71	- 6.0	650	65	0.32
150	20,650	137.7	.67	- 5.9	550	55	0.27
160	21,150	132.2	.63	- 5.5	500	50	0.24

Source: Basic data in Column 1 compiled by P.A. Briegleb, Pacific Northwest--Forest Range Experiment Station. All other figures derived from these data.

a/ Fully stocked acre, trees of same age; volumes are stem volumes, exclusive of bark and limbs, between stump and 4 inch top; stump height equals d.b.h. for trees up to 24 inches d.b.h., and 24 inches for larger trees. Figures rounded to nearest 50. Management at "natural" level--fire control, but no thinning, no fertilizing.

1/ d.b.h. means diameter breast high.

2/ Column 2 is column 1 divided by age; column 3 is column 2 as percent of column 1.

3/ Column 4 is difference between volumes in previous and curreng age in column 2.

4/ Column 5 is difference between volumes in previous and current ages in column 1.

5/ Column 6 is column 5 divided by 10, to put decade differences on an average annual basis.

6/ Column 7 is column 6 divided by column 1.

the stand is established; this might be from one to fifteen or more years after the harvest of the previous stand--one major management practice is to reduce this time as much as is economically rational.

There is no measured volume of harvestable wood on this site until the stand is about twenty years old; then the volume of harvestable wood in the stand rises rapidly, for a few years at a rising rate per year, then at a declining rate of growth (figure 1). It should be stressed again that these data relate to the volume of wood in the standing trees at each date; there is no allowance for wood growth on trees which have died or fallen or both, which might or might not have been salvaged. The data are for net, not for gross, growth of wood. In this kind of forest on this kind of site, under intensive management as much wood could have been salvaged in thinnings as would remain after perhaps sixty years of life. Such thinnnings, if skillfully done, would have little if any effect upon the volume of wood standing at any date--as some trees were taken out, the space they occupied, the sunlight they absorbed, and the moisture they took from the soil would be available for the remaining trees, and annual net growth of wood per acre would change but little. Such growth as did take place would be on fewer and hence larger stems. Thinnings are primarily a way of man's speeding up the natural selection and natural weeding out process and at the same time salvaging a great deal of useful wood.

Table 1 and figure 1 do not go beyond 160 years. Some natural stands of Douglas fir are as old as 400 years. Beyond 160 years, the trees might stand for a long time, probably growing slowly for a few decades longer, but it would be doubtful in this situation if the total standing volume would ever exceed 25,000 cubic

Figure 1. Volume of standing timber 5.0 inches or more d. b. h., by stand age, Douglas fir, site class II, site index 180.

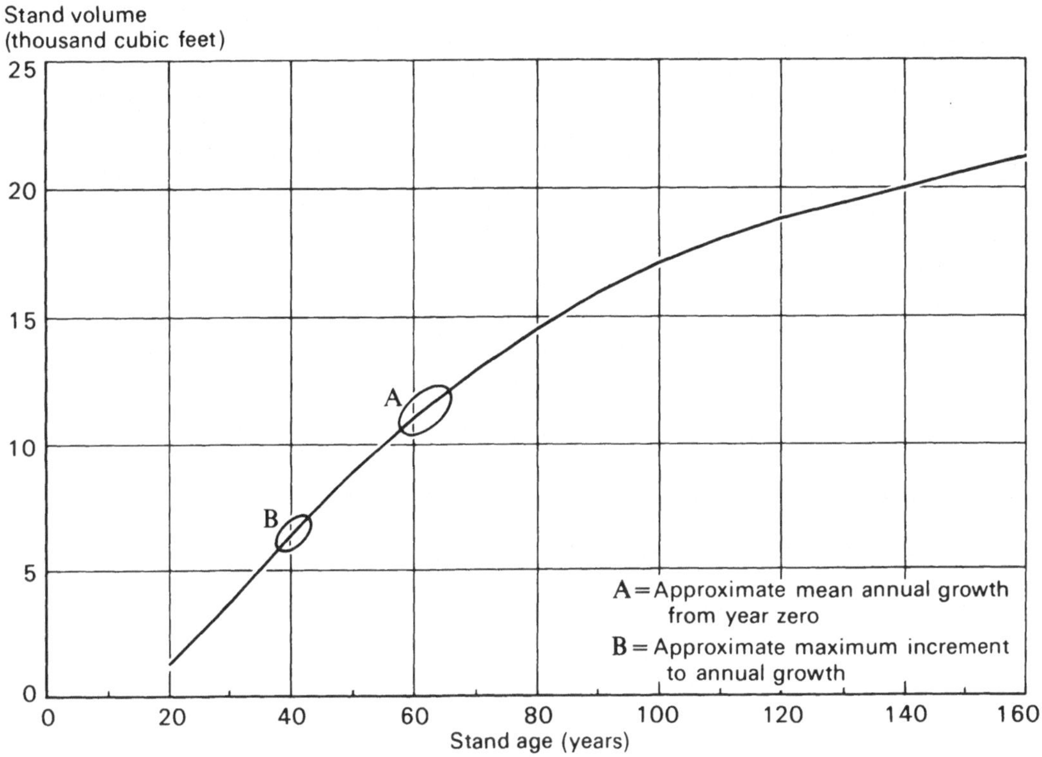

feet per acre. An old stand might succumb suddenly and dramatically, if struck by severe fire or severe storm, especially severe windstorm. If this occurred--as it has in the past on several sites--net growth in some year would become negative by the full amount of the standing volume. Another possible outcome of a very old Douglas fir stand is that it would be gradually replaced by hemlock, which can sprout and grow in the shade of the older trees. In this case, the volume of hemlock would rise, over time, in a fashion somewhat similar to the volume changes in the table and the figure for Douglas fir. The latter species would gradually or suddenly decline, under these circumstances, and its net standing volume per acre would almost surely decline as the older trees died. It is most unlikely, in the situation here under consideration, that the Douglas fir would succeed itself without some catastrophic clearing of the old trees from the site, since coastal Douglas fir is intolerant of its own shade throughout most of its range.

The data in table 1 and figure 1 are for one form of timber management, for one defined site, and one specified method of forest management. Although these data are basically a biological yield over time, they unavoidably include some technological, economic, and management factors. Volumes are for trees 5.0 inches in diameter; measurements of smaller trees would raise volumes while use of a larger cut-off size would lower volumes, for instance. The data are for good "natural" forestry, wherein the site is fully stocked naturally or by planting and fully protected from fire, but with few if any other specific forest management practices during the life of the trees. Many other curves could be drawn, for other forms of forest management. By varying the management practices a different response curve would result from forestry on this site. That is, special measures might be taken to insure the optimum spacing of seedlings and to insure a complete stand (not allowing the vagaries of natural reseeding to produce a sparse or an overful, or an irregular stand); the seedlings might be grown from specially selected seed, from superior trees well adapted to growth in this location; fertilization at appropriate intervals and by needed nutrients could stimulate growth; and thinnings at appropriate dates and intensities could produce a lot of usable wood and at the same time increase the average size (but probably not the total volume) of the trees surviving to harvest date. Each of these measures could significantly affect the growth rate; their combination, or their synergistic effect, would be greater than the sum of their individual effects. That is, fertilizing a thin stand may add little to net growth, or establishing an optimum spacing will be less effective if the soil fertility will not yield a higher rate of growth.

These management practices to increase wood growth would shift the curve in figure 1 both vertically and horizontally. That is, at any age the volume of standing timber would be greater, or the age at which a given volume of standing timber would be attained would be less or both. The curve under intensive management might lie to the left and above the curve for natural management in figure 1;in addition, its shape might be different--more steeply rising during the critical growing years.

Somewhat similar curves exist for other tree species and for other site conditions. The situation is more complicated where there are trees of mixed ages, although frequently a forest of mixed ages actually consists of small spots of trees of largely the same age. In actual forest management, an owner/manager would want the most accurate age-growth curves he could get at a reasonable cost, for his site and for the stand now on the site, if it is now forested. My concern is not to provide detailed data for many practical forest situations in the United States, but rather to illustrate particular relationships.

For the site in table 1, the average annual growth rate (column 2 in table 1) during the period from the date of stand establishment until the date of measurement reaches a peak at about sixty years (the "mean annual increment", in foresters' language), after which it declines very slowly in terms of cubic feet of wood grown annually (figure 2). The average annual growth of wood, as a percentage of the volume of wood standing at each date (column 3 in table 1), declines steadily from twenty years onward. The volume of net annual increment of growth in each decade, (colume 6 in table 1), reaches a peak at about forty years, after which it falls sharply, while this volume as a percentage of the volume of standing timber, declines sharply from the age of twenty years onward (as shown in colume 7 of this table).

In growth curves based on perfect data, the point of maximum average growth should arise at the point where the marginal growth curve crosses the average growth curve. It is easy, by formula or by arithmetic example, to show that this is the case. In table 1 and in figure 2, this seems now to be exactly the case, but the reason lies in the fact that the stand measurements are taken at ten year intervals, not annually or at relatively close dates, and the "marginal" curves are not true margins, but rather averages for a small period of years preceding the date.

Figure 2. Annual growth by age and in relationship to standing timber volume, Douglas fir, site class II, site index 180.

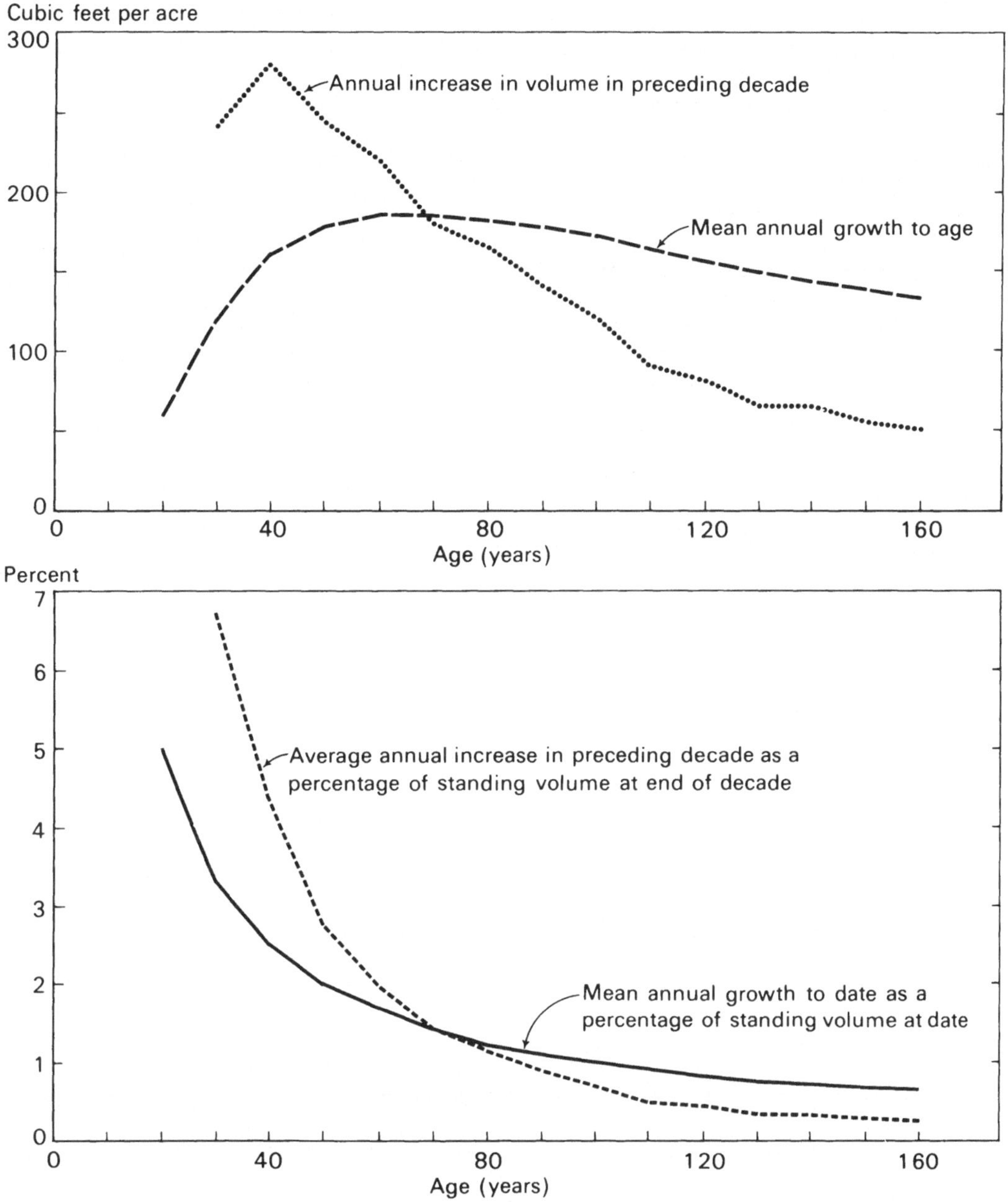

As the stand gets older, there are fewer and fewer trees, each of which is larger and larger, until both total wood volume per acre and size of individual trees is impressive indeed. At age 60, the trees on the site described in table 1 and under those management conditions would mostly range from 12 to 16 inches in diameter; by age one hundred the remaining trees would mostly range from 20 to 22 inches in diameter. But the annual growth of new wood per acre slows down, in this case having reached a peak somewhere around forty years of age, and the growth as a percentage of the volume of the standing trees declines very sharply from twenty years onward. If we remember that the standing trees are part of the productive factory (land is important too) and the growth of new wood is the output, then it is apparent that the volume of output per unit of standing timber volume falls greatly as trees get older, however impressive the old trees may look.

These data raise a question: what kind of wood is most profitable for the owner/manager to grow? The matter of what kind of trees he wants to grow, or to maintain, for aesthetic or other nonharvest reasons involves other issues, which we do not consider in this report. If the owner/manager thinks that the production of large timber--8" X 30" X 24' beams, as an extreme, or even 2" X 12" X 16' planks--will be the most profitable, then of course the trees must be big enough to yield such beams and planks. If he wants to produce 2" X 4" X 8' studding--a very common lumber mill output these days, to supply a basic item in housing and other construction--then he can do so from very much smaller trees. Moreover, increasingly the big beams of today are constructed by lamination of several layers of 2" X 4"'s. If he wants to produce pulpwood, then of course size of the stems is not basic as far as the production process is concerned, although the volume of wood is a critical fact.

The growth data in table 1 and in figures 1 and 2 show that the annual growth of wood per acre does not rise to a sharp peak and then sharply fall away. A tree is not a soufflé which has only one brief moment of perfection, at which it must be consumed. In practice, it is often hard to measure the growth rate on a particular tract of land with high exactitude. In any event, the annual growth rate changes rather slowly over several years near the peak. One cannot say, from the data in table 1, that this forest reaches its maximum annual growth rate at 59 years, 6 months, and 14 days; one cannot even say with confidence that it reaches the maximum at 59 years, as contrasted with 58 or 60 years; one can say, somewhere around 60 years of age, average annual growth rate over the life of the stand reaches a maximum, but that this may be reached on this site and under these general conditions as early as 50 years or as late as 70 years; one might, with good data, be able to say the maximum was reached between 55 and 65 years of age; and one can certainly say, in the case of these data, that the maximum annual rate of growth was not reached as early as 40 years of age nor yet as late as 90 years of age.

With growth rates changing slowly over a span of years, any timing of harvest which seeks to take advantage of a particular growth relationship can be varied considerably without significant consequences on growth rate. If the growth curve under intensive management is steepened, this might well lengthen the span of years over which harvest could be undertaken without serious effect on growth rate.

Timing of Timber Harvest

Harvest is one of the most important management actions in forestry; a decision to harvest or not to harvest a given stand of timber is a major management decision point on juncture in the history of a forest stand. The way in which the harvest is carried out and the resulting product is sold greatly affects, even determines, the income from the forest. It also greatly affects the future output of the forest. Foresters have understandably devoted great attention to the matter of timber harvest. How shall a forest owner/manager decide when to harvest a particular stand of timber or decide how much timber he shall harvest in a particular year? This is the subject of this section.

Three related but separate concepts should be distinguished:

A. Biological relationships of age, growth, and maturity of trees, and analyses based upon such information;

B. Economic analyses, especially considerations of financial maturity; and

C. Production functions, or the relation of timber growth to inputs of productive resources, which determine how much and what kinds of timber will be grown for the next harvest.

These three concepts frequently lead to others, such as sustained yield, allowable cut, even flow, sustained yield effect, and others, but I postpone the discussion of these latter relationships or concepts until much later, when I consider the special problems and opportunities of the national forests. But each of these three concepts must be explored somewhat more fully.

A. Biological analysis is based upon the kind of data presented in table 1 and in figures 1 and 2. It includes such matters as a consideration of "maturity" in trees and in stands of timber. Should the forester seek maximum volume per tree, per acre of forest, or per year of tree growth at the time of harvest? Most foresters have opted for maximum mean annual increment (MAI) in wood growth, a point which lies within area A of figure 1. If timber stands are harvested at this age, and if another cycle of timber growing of the same age-growth relationships follows, from which trees are again harvested at the same relative age, then the volume of wood grown per acre on this site is maximized over a long period of time, and hence is maximized on an average annual basis. The determination of the age of maximum MAI is subject to the uncertainties discussed previously, and forest management to achieve maximum MAI is subject to various uncertainties of climate, disease, pests, and other biological and physical hazards. Maximization of MAI ignores all considerations of costs of forest management, of markets for the wood grown, of wood processing, and of prices for wood, including all trends and short term variability in each of these factors. As generally applied by foresters, maximization of MAI assumes, sometimes implicitly, constant technology of wood conversion from tree to finished product, including constant labor input per unit of wood processed. Given these assumptions, maximization of MAI provides community stability of employment and income.

Maximization of MAI has been widely adopted and used by foresters as a management principle. This concept is most easily applied to the "fully regulated forest" as conceived by the forester, wherein there exists an equal distrubtion of timber volume, of timber acreage if all land has equal productive capacity, and of annual growth rate among the various age classes of the forest. There are great difficulties of applying this concept to the kind of old growth forests which dominated American forestry for so long and which still dominate the national forests. Foresters have devised many methods for application of the concept of maximizing MAI to the old growth Forests.

If the objective of forest management is to maximize the volume of annual wood production without concern for costs, then harvest at the maximum MAI is the proper point and it is irrational to hold the trees beyond area A on figure 1. Many foresters, motivated by "silvicultural considerations" or out of a deep professional commitment or bias for large trees have tended to postpone timber harvest until relatively advanced ages of the trees. Hershleifer has characterized the age-yield curve beyond area A on Figure 1 up to the point of maximum growth per acre, as the "first order of absurdity", since in this range the marginal rate of wood growth is negative; and he characterizes the holding of trees beyond the age of maximum total growth per acre as the range of "second order absurdity" since the rate of total growth per acre is now negative [3]. A stand of mature old growth timber, where rot, decay, windthrow, insect and diseases, and other hazards took away more timber volume annually than was grown, hence where net growth was negative, would fall in his second order of absurdity. Such stands are not unknown in the United States today.

B. Economic analysis of timber growing and harvest is concerned with costs and returns of timber growing, particularly costs as related to the age of the trees, in order to decide on the optimum timber harvest date. For well over one hundred years, some forest economists have been concerned with the most profitable date of timber harvest; some of their approaches and methods or analysis are discussed briefly in Appendix A. for those who wish to pursue the subject on a more sophisticated level. The introduction of costs into the decision process introduces many complexities, and the answer is no longer as simple as it seemed when the maximization of wood production was the basic criterion. The costs of timber growing are of three basically quite different kinds:

1. Inputs of capital, labor, and materials during the production process for seed stock grown in nurseries or for seeds, labor for planting, fire protection and control activities, fertilizer, and the like. Under intensive forestry, the cost of such items is much larger than the cost of similar items under extensive forestry. Some of these costs occur at date of stand establishment and some occur more or less evenly throughout the life of the stand.

2. Interest on the capital that might be realizable from harvest of the stand. This cost begins at the age when the stand could be harvested to produce some income, which could then be invested elsewhere to earn an income which is thus properly a charge against the timber production if the stand is allowed to grow. This cost rises during the rest of the life of the stand as the timber volume, and hence presumably the timber value, of the stand rises. While there should be a cost of this kind before the stand reaches a minimum harvestable age, it is often not readily measurable; the young growing stand may have value but it is a value based upon an anticipated later harvest value. It is a value which the owner could convert into liquid capital only by selling the

property; and the market for immature stands is so imperfect in many areas, in the economist's sense of that term, that sale value of the young trees is often indeterminate.

3. Rent on the bare or unforested land, based on the best alternative value of the land for other uses or the rent which may be earned in the next timber rotation. If the forest land is incapable of any other use, then the land rent is zero except as rent may be earned from the forestry operation itself. For many sites, the earnings from the next rotation of timber growing are highly speculative or dubious, partly because it is so far in the future, and it may be negative.

Mason Gaffney in a classic paper has developed the concept of financial maturity as applied to timber (and other assets)[4]. As he well points out, the factors making for prompt harvest of a timber stand are basically two: (1) an urgency to receive the income available from the timber harvest, so that it may be used or invested elsewhere (or into a new timber rotation); and (2) an urgency to free the site for further timber growing. The first of these factors is reasonably well measured by the interest, at the best alternative investment, on the net sum realizable from the timber harvest. The second factor is best measured by rent on the bare land. This is often indeterminate in a market, since the market for land is often imperfect. Gaffney proposes using the rent which may be earned on the next production cycle of timber on this land; but this in turn is often dependent upon the age and output of the forest for the next cycle. This requires a simultaneous determination of rent and of length of timber rotation.

Against these forces making for timber harvest is the net income which may be produced by letting the trees stand and grow for a longer period.

It would be easily possible to make a benefit-cost, or income-expense, analysis or budget year by year throughout the lifetime of the timber stand to find the year when the cumulative margin of income over costs was the largest. This would be the age of financial maturity of the stand. It would invariably be younger than the age of MAI, by amounts which would vary according to a number of factors. Instead of this year by year calculation, economists have developed different formulas which include the various kinds of costs in different ways, and include an increase in timber growth and value each over time, which can be solved to determine the optimum age of harvest. This type of approach has all the uncertainties that afflict the biological approach, as noted earlier, and in addition has uncertainties of future costs, future timber prices, future interest rates, and others. Since the economic optimum relates to income above all costs, there may well be a question of which item of input whose net return should be maximized. As Gaffney wisely says (p. 57): "The choice of what to maximize is a matter of judgment. In general, one should maximize the return to the input that is hardest to evaluate in existing markets or other alternative uses. As a rule, that is the site. But small firms constricted by stringent credit rationing may want to maximize i (inter est rate)."

Economic theorizing about optimum data of timber harvest, reported in Appendix A like all good theory, is an abstraction from reality. That is, simplifying assumptions are made to measure the basic effect of the major variables studied. The resultant theory is basic to any analysis of practical problems but it alone may not be sufficient to enable the forest owner/manager to deal with the numerous complex situations that may be encountered. The theoretical economic approach may pass too lightly over some difficulties, may ignore others, or may be unaware of many forest situations and relationships, or may assume away many practical problems. In particular, it may be, and has been in some instances, deficient in one or more of the following ways:

1. It implicitly may assume a precision to the age-growth curves of forest stands which they do not, and cannot, possess. At the best, such curves are approximate. Even if reasonably accurate, they cannot pinpoint an exact date at which average wood growth or marginal wood growth is at a maximum, but rather indicate a range within which the exact date probably falls. Any analysis based on such curves must equally be approximate, not precise.

2. Some of the analyses do not include differences in forest site productivity. Some read as if, to paraphrase Gertrude Stein, a forest is a forest is a forest, and that which is applicable to one forest is equally applicable to another. In fact, of course, commercial forests in the United States differ greatly in physical productivity: site class I will produce more than 165 cubic feet of wood annually, while site class V will produce only 20 to 50 cubic feet of wood annually, and extremes of actual forest situations are greater than these class averages. Economic variation is vastly greater than the physical variation; costs tend, to a large extent, to be a function of forest area, while income is directly a function of volume of wood grown. Marty has calculated that, under the same assumptions of future timber prices, some forest types on some site classes will make a percentage return on capital

several times as great as other site classes for the same forest types [5]. For the manager of large forest properties, which will almost surely include some land of widely differing productivity, efficient timber management calls for quite different programs on different sites. The economic maturity approach is fully applicable, and in practice sometimes has been applied to forest sites of different productivity.

3. Some of the analyses give little or no attention to income from wood production intermediate between stand establishment and final harvest. Gaffney shows how this can be worked into his analysis, but virtually all of his discussion is in terms of a lump-sum, one-time-only, beginning-of-cycle cost.

4. All the economic approaches to measuring optimum harvest data rely basically, and heavily, upon an interest rate, either as an externally-determined rate or as a rate to be maximized internally and compared with some external rate. Yet none of them give any real guide to the forest manager to aid his choice of the appropriate interest rate. Most economists would agree that the proper rate would be different, depending upon the situation of the particular forest owner/manager. Perhaps no other single factor will affect the decision about timber harvest as much as the interest rate does.

5. Many economic studies either ignore the market for wood or assume away the problems of marketing. Often they ignore the effect of the volume of harvest upon the market price. For the small landowner, this is appropriate, but for a large landowner, either private or public, it may be unrealistic. For a very large enterprise such as the national forests, the effect of timing of timber harvest upon market price for

timber cannot be ignored. Some of the more sophisticated economic models explicitly introduce market price effects of the timber harvest schedule [7].

6. Most of these economic studies ignore the problems of processing the timber harvested. They assume that a market exists or they ignore the problem. There is generally no explicit consideration to the fact that the logs taken from the forest must proceed through a relatively extensive processing, transportation, and marketing process before they are usable, and hence valuable, to a final consumer. There often seems to be an assumption that an adequate processing capacity exists to handle any volume of logs that might be harvested as a result of the application of any formula, and there is no concern about the optimum scale of processing plant or the optimum rate of timber harvest to realize economies in timber processing. Again, these assumptions make little or no difference to the small forest owner, whose volume is too low to have any effect upon processing and marketing, but they may be crucial to a large integrated timber firm or to a large public enterprise such as the national forests.

C. A production function is the relationship between inputs of productive factors and output of the product. For forestry, it is the relation of land, standing forest, annual inputs such as fertilizer and fire control, interest on capital, and managerial input to the output of timber. As such, it is not strictly a harvest matter, and certainly not a matter of harvest timing. But the production function determines what will be available for the next harvest, and to this extent is part of the harvest process. The conditions of the harvest in turn may greatly affect the production function, particularly as the condition of the site is affected by the timber harvest.

The growth of timber is always a function of all the inputs described above. In "natural" forest management, all the inputs except the land are at a low level. The trees fight their own battle for establishment, survival, and growth. A great many of the economic analyses referred to above assume, sometimes implicitly, "natural" or minimum forest management. Many more intensive methods of forestry are possible, some of which are discussed in a later section of this report. They all involve greater costs and they involve, or should involve, larger gross returns. Unless they promise larger net returns also, such intensive methods are not profitable. The more intensive the management, the higher the cost of inputs, the greater the value of outputs, and the greater the economic uncertainties arising out of possible future changes in costs and prices, as well as the biological uncertainties previously discussed.

The options for different forms of forest management are greater immediately after a harvest than at any other time in the life cycle of the timber stand. Then it is possible to affect, if not fully decide, the genetic nature of the next timber stand, the time interval between harvest and stand regeneration, the spacing of the trees in the new stand, and some other factors, each of which individually and all of which collectively to a far greater degree influence timber growth rate. But a production function exists at every stage of stand growth. Harvest of part of a stand is often a major production input into the further growth of the stand.

Determination of the optimum intensity of forest management is specifically an economic matter, although of course it depends upon the biological relationships between input factors and output. Determination of optimum age of harvest in turn depends upon the optimum intensity of management.

In very general terms, there are six major alternative dates or stages in tree growth at which timber harvest might occur:

1. Never. This is the case under present law in national parks, wilderness areas, and some other reservations. A private landowner might prefer to look at his forest in an unharvested state, hence would exclude all harvest. The next owner of the same forest, of course, might follow a different course of management. But, as we shall point out in a later section, no harvest might also be economically optimum for public lands where there was a legislative or ideological commitment to continued timber management but at costs which exceed the values that can be created.

2. When the tree falls or when major falldown seems imminent. This might make sense in a park or other reserved public area, or on a private holding, where timber harvest was normally eonstrained or prohibited, if the harvest not only salvaged some values but reduced or eliminated hazards to personal safety of visitors. Even when a tree falls, often there is some salvagable value. Ecologists, silviculturalists, entomologists, and others can often predict with considerable accuracy the probable demise of a tree. Its removal when early fall seems probable might increase its salvage value as well as reduce the risk it would fall and injure someone. Timber harvest under these conditions would normally be of very old and highly decadent timber stands.

3. Salvage of dead, down, or dying trees of any age. Climatic, fire, and biological hazards kill or seriously cripple many stands of timber long before they run out their life span as suggested in figure 1. A substantial part of all timber reaching the market each year in the United States is salvage timber. Good forest management should expect events which will make salvage economically and ecologically desirable and should be prepared to move promptly when salvage is necessary. There is no clear line between this type of timber harvest and the foregoing, but salvage is (in my terms) more an economic matter, on forests managed for timber growth and harvest, while harvest of threatened trees is more one aspect of management of the forest for values other than timber harvest.

4. At maximum timber volume per acre. Some definitions of tree "maturity" and some discussions of public timber management seem to assume that harvest should occur only when trees have reached their maximum attainable size. Such harvest would be of very old trees, but not older than the harvests currently occurring on some national forests where net growth per acre is already negative. Again, there is not a clear line between harvests on this basis and those of the two preceding bases. If trees are decadent and harvest is undertaken to reduce the hazard of their falling to someone's injury, they may also be at or close to their maximum size. While conditions requiring salvage may strike trees of any age, the older trees are much more susceptible to some types of hazard, notably such biologic hazards as disease and insect attack.

5. At maximum MAI. As noted earlier, this is the age of harvest chosen by many foresters, who defend their choice--to the extent they

regard defense as necessary--on the grounds of maximizing timber production from a given area of land. MAI takes on a new significance in light of the National Forest Management Act of 1976, which states in section (6(m)): "The Secretary shall establish (1) standards to insure that, prior to harvest, stands of trees throughout the National Forest System shall generally have reached the culmination of mean annual increment of growth..." The probable effect of the Act on the management of national forests is discussed in a later section of this report; determination of the age of maximum MAI will not be easy under this act. Not only is there the general imprecision of measuring growth from year to year, and the dependence upon what sizes (and sometimes types) of timber are to be counted as "growth", which have been discussed above, but MAI is also affected by the intensity of forest management. Intensive management will almost certainly shift the age of maximum MAI to an earlier age, perhaps to quite a lot earlier. One may safely predict many arguments over the application of this part of the 1976 act.

6. At financial maturity. This is the age of harvest when the relationship between costs of timber production and values of timber grown are the most favorable over the lifetime of more than one stand of timber on a particular site. Its characteristics and its problems of measurement have been discussed above. Even if financial maturity is the general goal, there may be a need in practice to modify the exact date of harvest, as is discussed in the sections which follow.

Components of Timber Supply

Instead of talking about timber "supply", it is more enlightening to talk about "willingness to harvest" and about "willingness to grow" timber.

The owner/manager of any forest area (except one which has recently cleared of trees by harvest, fire, or some other action) typically faces a given stand of trees on the property. In the case of the national forests, the Forest Service and other parts of the executive and legislative branches face very large inherited volumes of mature standing timber Other forest owners may face different volumes of timber in their forest stand, but typically they will have some timber of harvestable age. When should all or part of the existent stand be harvested, how, and under what terms? Should harvest proceed quickly, or be postponed? This problem involves willingness of the owner/manager to harvest what he has. It is not unlike the willingness of a store owner to sell the goods he has on his shelves.

For a forest property that has been harvested recently or otherwise has lost its timber stand recently, a different question arises: is the owner/manager willing to invest funds to grow more timber on this site? If the land is not to be used for some other purpose, and regardless of its ownership, then some volume of trees will grow "naturally" in much of the United States. That is, in the absence of vigorous efforts to prevent tree growth, most presently forested sites in the United States will produce a new stand of trees. The new stand may differ in species from the old one, its density may be far from optimum, the rate of wood growth may be far below the "natural" potential, the quality of the logs may be low, and in other ways the new forest may be an unproductive one. Our concern here is not with this type of minimum natural forest regrowth, but with positive investment, including investment of managerial time and ability, to produce more and better wood from the site. The response of different land owners to this question does, or should,vary according to other aspects of their particular situations.

Although these two aspects of timber supply should be separated in any analysis of timber situations because they involve quite different factors, in practice the first type of response--willingness to harvest--gradually shades into the second one. When a forest owner harvests a particular stand of timber, or sells the standing stumpage to someone else who harvests it, shall he run through the saw every log which seems to produce a net value above the marginal costs of harvesting, processing and transporting its product? Or shall he leave some of the existent stand to grow for a later harvest, even though the logs are (perhaps marginally) profitable to harvest now? Should the owner take precautions himself, or insist upon them if he sells the standing timber, to insure that minimum damage is done to the soils on the site and to the young timber growing on it, not yet harvestable (even by the most inclusive definition)? Costs may thus be incurred, or income foregone, in harvest at one date, in the hopes or expectation that this increases timber volume and value at a later harvest or both--harvest gets translated into willingness to grow.

Two further comments may appropriately be made to add to the last point. In the first place, the appropriate decision will depend in large part upon the forest type and its growth habits; what is optimum age of harvest for an even-aged loblolly pine stand in the Southeast may not be optimum for a mixed age Ponderosa pine stand in the Northwest, and so on. Secondly, in this type of situation, sometimes the marginal costs of harvesting a log at the time of major harvest are much lower than the cost of harvesting the same log will be at a later date when it has grown larger. Roads constructed for the harvest of the larger logs will serve, often without any additional cost, as roads to harvest smaller logs; timber fellers can often fell small trees along with large ones, at lower marginal costs; likewise, skidding the logs out to a central loading point may involve some economies for the smaller logs; and, if the sawmill is a small portable one, as is often the case for the harvest of a small stand of timber, the marginal costs of its operation for additional logs, even if each is smaller, may be lower than its average costs. For all of these and perhaps for other reasons the harvests of stands of timber, especially from relatively small areas, have tended to push each harvest well down the scale to smaller logs, and to reduce consideration for the next harvest to a minimum.

Forest Management Situation Dominates Optimum Timber Management

The forest owner/manager always faces a specific management situation on his forest which dominates the choice of the timber harvest date. There are general principles of timber harvest which apply to all sites, but the specifics differ according to factors peculiar to each site or to each owner. There are a great many different forest management situations in the United States; elsewhere I have estimated that there are literally 4,000 different management situations among commercial forests [2]. There are some 210 species of trees with commercial value; these are generally grouped into some 20 forest types. Each of these has its particular growth requirements and outputs. There are also geographic differences, either states or regions; much of the data on forests recognize ten major regions. But regional and state data do not always differentiate important geographic differences, as between coastal plain and Pidemont in the Southeast. Five different site productivity classes are usually recognized for each forest type and each regional location. At least four major forest ownership classes must be recognized--national forests, other public forests, forest industry forests, and "other"private forests. The latter can appropriately be divided further. Sites also differ, at any given date, according to the degree of timber stocking, which may greatly influence an owner's willingness and ability to harvest timber at that date. Forests also differ according to insect and disease infestation, which may in turn affect the optimum timber harvest and timber management plan. Not all forest types are found in all regions, and some other possible classifications of timber management situations are absent or quite unimportant. But a consideration of all these variables results in my estimate of 4,000 timber management situations.

It is impossible, in a report such as this, to consider any large part of the 4,000 different forest management situations. Instead, three typical or important situations are considered, to show how the classical economic analysis of optimum date of timber harvest may be applied in actual management situations. The three situations chosen for analysis are: small private owners in general, vertically integrated forest industry owners, and national forests. For each, the analysis is first in terms of willingness to harvest existing stands of timber and secondly in terms of willingness to invest money to grow more timber.

There are perhaps as many as 4 million private forest owners in the United States. In total, they own nearly 60 percent of the total commercial forest land of the country. According to the available data (unfortunately 25 years old and perhaps even then not wholly accurate), over 85 percent of this area is in "small" (less than 5,000 acres) properties and nearly three-fourths in properties of less than 500 acres each. There are, obviously, many important differences in management situations among a group as large as this. Their dominant characteristic for the present analysis is that no small forest owner, with rare exceptions, possesses either wood processing facilities or ability to offer enough timber to affect market price--certainly not nationally or regionally, and only rarely locally. This class of forest owner must accept the market as "given" to him, and adjust his operations to it. As part of the market, the existence and the management practices of wood processors may be highly important, for they may affect greatly his ability to market what he would like to grow.

The forest industry firms own about 13 percent of the commercial forest land of the country. Their forests have, on the average, a somewhat greater potential productive capacity per acre than do the average forests of any other major ownership class. This is partly because the forest industry firms have chosen more of the highly productive sites but also because they have more largely avoided the low productivity sites. However, variability in productive capacity within each ownership group is more striking than are average differences between ownership classes. Forest industry firms, by definition, possess some timber processing capacity. There are about thirty major firms, vertically integrated from forest to final marketing ability. Some operate largely or wholly on their own timber, others buy much or most of their timber, but all have at least some timber of their own, and all have extensive processing capacity. Many have extensive marketing facilities, with trademarked products of various kinds.

The dominant fact about these vertically integrated firms, as far as the present analysis is concerned, is that wood growing must be considered as part of the wood-processing and market activity of the same firms. We shall show that the economics of wood processing and marketing are important, perhaps dominant, in their decisions about wood growing.

Our third example is the national forests, which contain about 18 percent of the commercial forest land of the country, and currently have about 50 percent of the standing softwood timber volume. These forests are thus very large in area and in volume; they are publicly, not privately owned; and they contain very large volumes of inherited old growth timber, much of which is at ages where the growth rate is far to the right of area A on figure 1. On some of these forests, net growth is zero, since rot and decay offset the relatively low rate of gross growth; and for some, net growth is minus. By usual economic standards, such forests are past the optimum harvesting age. Timber is sold from the national forests, to buyers who possess processing capacity, some of whom also have timber of their own. The national forests are not vertically integrated, since they possess no processing capacity, yet their production and sale of timber is dependent upon firms which do have processing capacity. The purchasers of timber from national forests have varying degrees of dependency of timber supply from the national forests; some are wholly dependent on this source, others only modestly dependent. The Forest Service, as the manager of the national forests, has always been much concerned with the effect of its harvest and sale policies upon the local communities where the timber is processed.

There are obviously many management situations intermediate between these three which we do examine, but the characteristics of those intermediate situations and of their management decisions can perhaps be inferred from the three which we do discuss.

Willingness of Small Private Owner to Harvest Existing Stand

Any small private forest owner might reasonably consider the possibility of harvesting some of this existing timber stand, if some of the trees are large enough to harvest and yield a positive return over the costs of harvest. An owner might face this alternative at any time in his history of ownership of a tract, but often a person who buys a tract of forest land for any reason is likely to consider the possibility of selling the merchantable trees in an effort to reduce his investment.

With so many small private forest owners in the country, clearly there are many different management situations facing some of them. The average owner in this category in 1970 (the latest year for which there are comprehensive data) had 1,340 cubic feet of growing stock per acre in his forest; this compares with 2,010 cubic feet per acre on the average forest industry forest, and 2,990 cubic feet per acre on the national forests [1]. While these figures may not be exactly comparable, since the forest types for each ownership class are not identical and since some types normally have greater volume per acre than do other types, they are nearly enough comparable to show that the typical small private forest is lightly stocked. Owners of this class have frequently been criticized by foresters for this low rate of stocking, the argument being that growth rates per acre will be low because stocking rates are low. If stocking rate is low because the stand is thin, there is merit to this argument; if the stocking rate is low because recent and frequent harvests have reduced the volume per acre, then it may well be that the small private owner is a better economist, though a poorer forester, than his critics and would-be advisors, since he has kept his investment low.

This average rate of stocking of the small private forests cannot be compared directly with figure 1, because the forest types and the site classes are different. However, if an age-volume curve appropriate for the site and forest type of the small private forests was available, the present rate of stocking would almost surely be to the left of area A on figure 1 and might even be to the left of area B. If one refers to the lower part of figure,2, the age and stocking rate might well lie on the left of these curves, where the marginal growth per year was a comparatively high percentage of the volume of standing timber. It is the probability of this relationship which underlies the earlier remark about these small private forest owners perhaps being good economists.

In considering whether or not to offer his timber for sale, the small private forest owner would be well-advised to use biological data on growth rates and an economic analysis of income and of costs from holding for further growth. He would need reliable growth rate data for his forest, and for many small owners it would not be easy to acquire the necessary data or to make the necessary calculations. Often he would have to make judgments from data assembled for other, hopefully similar, sites. He would also have to consider his costs of holding the stand for a few years, and thus calculate the balance between expected income and expected costs. If this balance were negative, then immediate offering for sale would be the most reasonable course, all else equal (circumstances under which all else might not be equal are discussed later). If the balance between expected income and expected costs were positive, then he would need to calculate the probable rate of return on the probably realizable capital from sale of timber. This would then have to be compared to the rate of return he could earn, in investments actually open to him, with differences in risk of loss being also taken into account.

But the small forest owner should also consider several personal factors, not related to the forest as such. What would an early sale do to his income tax rate? At what future date would the income tax impact be the lowest? What is his need for income now? Is there any personal advantage in holding timber as a savings account, instead of cashing in its value and investing the returns elsewhere? For instance, many a family might find it possible to hold a growing stand of timber against the day when one or more children was ready for college, when they would otherwise spend the cash income from more frequent timber harvests for various consumption items, such as a new car. His need for cash and the possibility of getting it by timber sale might be the compelling reason for sale.

The foregoing discussion is highly relevant to the small forest owner's willingness to sell some of his standing timber, but it may have nothing to do with an actual sale. The latter requires a buyer. The local market for stumpage or standing timber is typically imperfect, in the economist's meaning of that term. There is often but one or a few buyers, and sometimes there is none. There may be a pulpmill within reasonable transportation distance, which stands ready at all times or nearly at all times to buy pulp wood of defined quality. The prices for pulpwood are typically fairly steady, having risen gradually over the past several years but not fluctuating significantly around the trend on a short-term basis. The paper companies have typically set prices for pulpwood to draw forth desired volumes of supply over a considerable period of time, supplementing the volume purchased by wood grown on lands owned or leased by the paper company (as will be discussed in a later section). The small forest owner facing this kind of a situation at least knows what he can sell, and approximately for how much, more or less whenever he wants to sell--assuming that he can get a harvesting crew at a reasonable rate. The price may or may not be highly profitable, and the chances for a killing when the market is booming may be lacking, but the reliability of a market is highly important.

The owner of a small forest who produces sawlogs may also have a permanent processing mill as a potential buyer within a reasonable transportation distance. Even if he does, the price of stumpage is much more variable than the price of pulpwood. Figure 3 shows prices actually received for timber sold by the Department of Natural Resources of the State of Washington, from July 1959 to April 1976 [8]. These data do not relate to a uniform species mix of logs sold, nor to a single location, but they do demonstrate to volatility in the market for sawlog stumpage. While this situation may be somewhat unusual, the prices are those actually realized in sales, not some price quotations which, for stumpage, are often not fully reliable. Timber sold in 1966 would have returned substantially less in this region than if sold in 1969 or 1970; and if sold in the latter years, substantially less than in 1973 or 1974. Even disregarding those major savings in stumpage price, there were substantial variations in price within some years. A small forest owner might not be able to get as good prices as did the Department of Natural Resources, but he might have done proportionately as well or even better in the high price years, when demand was active, than in the low price years.

A strictly economic analysis of optimum date of timber sale, based on growth rate and normal costs and longterm price trends, is useless in a situation such as figure 3. An analysis based on biological age-growth curves which assumed constant stumpage prices (as most such analyses do) would have been even worse. No conceivable fall-off in rate of timber

Figure 3. Average stumpage prices for all timber sold by Department of Natural Resources, State of Washington, July 1959 to April 1976.

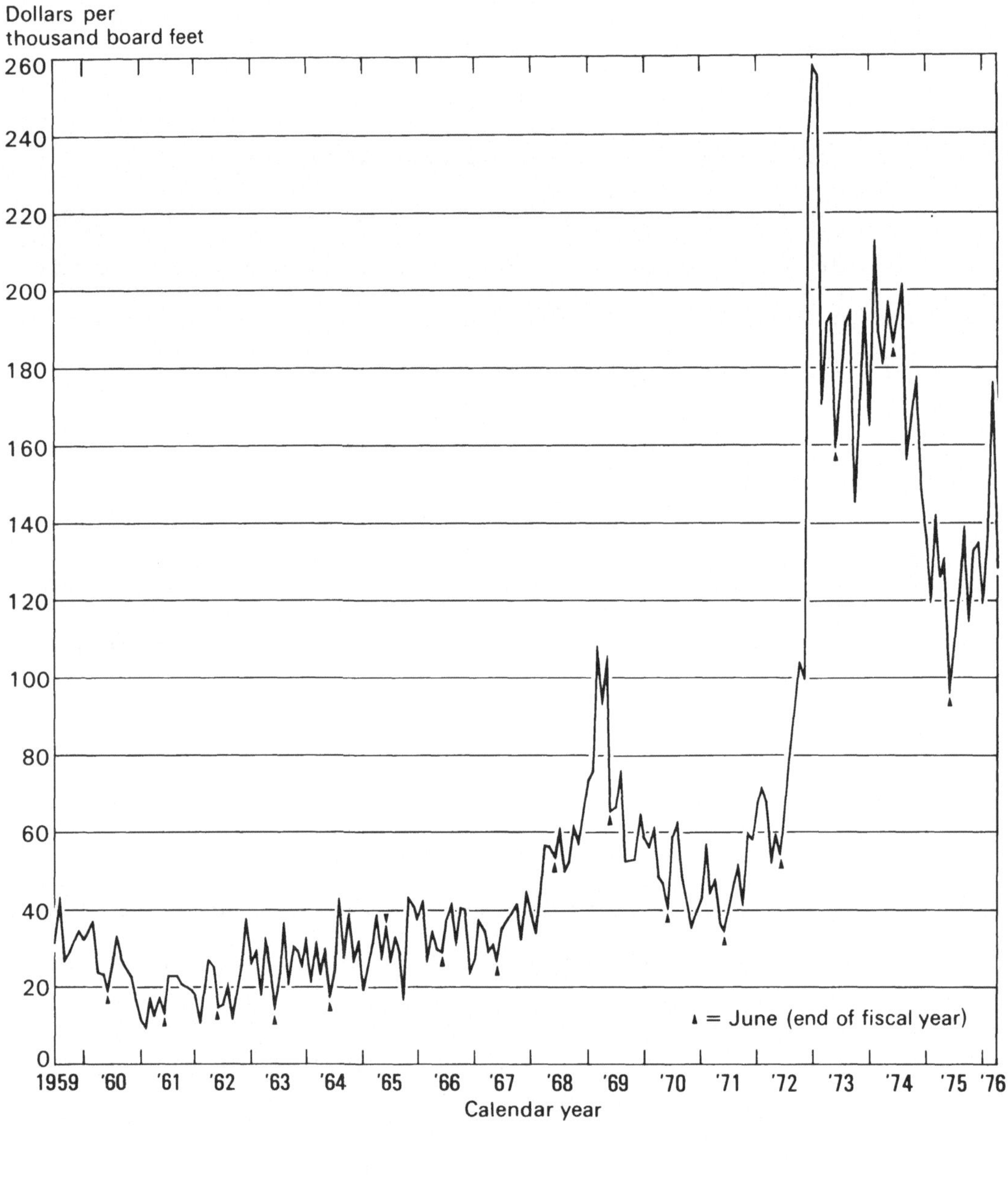

growth or no possible rise in costs between 1966 and 1976 could have offset the tremendous increase in stumpage prices between those years. It should be added, in candor, that no means exist for a fool-proof projection of timber prices a few years of even a few months ahead. But any increase in knowledge here will far more than equal any gains from analysis of optimum age of timber harvest, and any effort expended by the forest owner/manager to get the best realizable price for his logs will be much more remunerative to him than similar effort expended in growing logs.

The small forest owner is more commonly dependent upon the operator of a small portable sawmill, who brings his mill to the forest, for a market for his stumpage. The forest owner gets, or seeks, an offer for his timber from such a small processor. There may be some economies of scale here, even for a mill of the same (very small) size; that is, a larger volume to saw means a longer run, and hence a larger volume of lumber against which to charge the necessary costs of moving the mill. Should the forest owner sell now, with a relatively small volume to harvest, or wait a few years, in the hope that a larger volume will enable a small processor to pay somewhat more per unit of volume for it? The anwser may depend in part upon the kind of wood the forest owner has for sale. Woods of unusual value, as white oak of whiskey barrel grade, or high class walnut, may attract a good buyer even when in very small volume, when ordinary oaks, hickories, maples, and beeches would not. The owner's need for cash now might also be determinative.

More important than the possible economies of processing are the potentials of competition in the bidding for timber, as far as the small forest owner is concerned. A single bid for his timber will almost surely be less than the full value of the timber; the typical small sawmill operator exists in part because he is able to buy cheaply. Many small forest owners lack any basis for comparison; a buyer comes along, offers the owner a few hundred dollars for all the merchantable trees on his tract. The owner may have sold no timber for some years; especially in recent years when timber prices have risen so much, he may be completely out of touch with the market. The proferred money looks attractive, and he accepts it without real effort to find out if a better offer is possible. An old and sound rule is: don't sell unless you have three offers to choose among. It may well be that insistence on this rule means a lost sale, because no alternative buyer is on the scene. But if the timber can be held a few years, without serious or any physical loss in volume, possibly with considerable increase in volume, and without personal hardship to the forest owner, then the odds are very good that, sooner or later, a much more rewarding price can be obtained.

At any rate, the willingness of the small private forest owner to sell part of his standing volume will be, and should be, more influenced by market possibilities than by any other single factor.

Willingness of Small Private Forest Owner to Invest in Growing Wood

When we come to a consideration of the willingness of a small private forest owner to invest in the growing of wood, an important distinction must be made between the man who already owns a tract of forested or potentially forested land, and the man who is contemplating the purchase of land for this purpose. The former is faced with certain specifics of his site--its capacity to grow wood annually, its location with respect to markets, its location with respect to roads, its location with respect to where he lives, and other factors. The man who is contemplating the buying of a forest tract has much more flexibility in these matters. It is true that the man who owns a tract could sell it, and buy another, but there are many obstacles to this process. The market for land is imperfect, and he may have great difficulty in getting a price for his land which is as high, relative to its income-producing capacity, as he has to pay for the land he buys. There are substantial transfer costs involved in buying and selling land.

For either situation, a basic consideration is the owner's or the potential owner's objectives in forest land management. Is he seeking to obtain a satisfactory return, or even something higher, from his investment in the land and from his management of it, or does he have in mind some other set of objectives? Does he, for instance, look upon the land as a chance to make a large capital gain from a rising price of land? Or is he interested in the forest as a place for his personal recreation? Or, if he has inherited it, does it have personal value because of memories of earlier episodes in his life? These latter considerations all boil down to one basic question: how long will he continue to retain ownership of the land irrespective of its earnings from timber growth? Will he consider the whole forest land ownership and management as one economic package which must return an income competitive (or better) with any other investment of his capital and time? Or will he continue to own the land irrespective of income, in which case timber management must only return more than the <u>extra</u> costs incurred to grow more and better trees or both?

The answers to these questions are critical to the analysis of the forest owner's willingness to grow wood.

Irrespective of the owner's objectives in timber land ownership and management, the biological data on rate of tree growth and the traditional economic analysis of income in relation to costs, or more properly of marginal income to marginal costs, including the consideration of the optimum date of timber harvest, are basic to his forest management decisions. If his objectives are not to maximize income, or even to maximize income above extra costs of timber growing, then of course he might choose to retain trees far beyond their optimum harvest date. But if he is concerned with income, then alternative systems of forest management, including different degrees of intensity of timber management, and including optimum date of timber harvest, are critical. For these, the usual economic analysis is highly valuable, even fully adequate, as a basis for decision-making about investment in timber growing.

Even for a consideration of willingness to grow (as contrasted with a willingness to harvest) timber, market prospects at the probable date of future harvest are highly important and may be critical. How many possible buyers may there be for his timber? How much, if any, competition is there likely to be among buyers? What probable price differentials will there be between large and small size logs, and between high quality and ordinary quality logs, and among tree species which he might grow? Admittedly, it is difficult to have a reasonable degree of assurance on these points now, for a harvest which will be some years in the future, yet these questions are inescapable. When the time comes when a harvest must be considered carefully, the earlier factors affecting willingness to harvest become dominant.

For the United States as a whole, the purchase of a relatively small tract of forest or potential forest land at market prices and managing it for wood production do not offer very good prospects for as high a return on capital as might be earned elsewhere. Forest land prices have too often been bid up because of the forests recreational value to offer good returns to capital from timber growing. Small owners face many disadvantages in selling their output at good prices and in managing their properties efficiently. For the best timber growing sites, where there are good markets for the wood that can be grown, where costs of forest land ownership and of timber growing are relatively low, and where other factors are favorable, modestly high returns from timber growing are possible. Even then, it is doubtful that returns will be especially high, considering risks and the general illiquidity of investment in forest land.

If the forest land will be owned irrespective of income, and if the costs of land ownership are thus born by the other objectives of ownership (such as recreation) so that the forest enterprise must earn a competitive return only on those extra costs of timber growing, then the outlook is a good deal better. In this case, a satisfactory return may depend on keeping costs, including the interest which would be earned by harvesting the timber stand and investing the capital elsewhere, down as low as reasonably possible. Economic timber management under these conditions often means harvesting fairly frequently and to levels of timber stocking which the professional forester regards as undesirable. As we have noted, many of the small private timber owners have been more realistic in this regard than have been their advisers. Faced with an enterprise which cannot make satisfactory returns on all its parts, faced with an uncertainty of continued ownership, and desirous of some returns to make his extra investments at least tolerable, the small forest owner often follows practices which the forester concerned with "good forestry" finds abhorrent.

Willingness of a Vertically Integrated Forest Industry Firm to Harvest Its Timber

A vertically integrated forest industry firm, by definition, owns at least part of the timber which it processes and markets, and also by definition processes most or all of its own timber. The reasons why such a firm decides to grow any part of its needed timber supply, instead of purchasing all of it from other forest owners, will be considered in the following section. At this point, we assume (and this closely corresponds to reality) that the firm has the capacity to grow at least half as much timber as it has capacity to process. We assume it has a plant or plants of given character and output capacity, in a fixed location, and that it has a marketing establishment built up over a period of time to sell what it processes. Our concern in this section then is what factors affect the willingness of such a firm to harvest some of the timber it owns?

An answer to this question must start with a brief consideration of the characteristics of the vertically integrated forest industry firm. As noted, it includes capacity to grow timber, to process it, and to market it; in these characteristics it resembles an integrated petroleum industry firm which has oil fields, refineries, and service stations, or an integrated steel industry firm which owns ore deposits, steel mills, and marketing structure. The economic advantages (and disadvantages) of vertically integrated forms have been considered in economic literature and need not be repeated here. An examination of the annual reports of most of the major integrated forest industry firms leads me to the conclusion that the cost of wood as taken from the forest is of the order of 15 percent of the total costs of processing and marketing the finished products from a vertically integrated forest industry firm. Many firms seem to carry timber inventory at original acquisition cost rather than at present value, so a more realistic timber cost may be larger than 15 percent. Regardless of the exact figure it seems evident that timber in the woods is only one cost for vertically integrated forest firms.

As with nearly all vertically integrated firms, there are substantial economies in keeping a vertically integrated forest industry firm operating at or close to full capacity. A great many of the costs of such a firm are overhead costs, not responsive to short run changes in volume. This is true for marketing costs, including advertising costs, as well as for plant production operations. High overhead costs mean relatively low marginal costs. There are considerable economies in keeping a sawmill or a plywood plant operating at or near capacity, although operations can be reduced to only some days per week or can be increased by double shifts if the economic advantages are strong enough. The economies of continuous operation at full capacity are especially great for a pulpwood and paper mill; the cost of shutting down, and starting up again, are simply too great to permit short run flexibility in output. At all costs, wood must be available to keep the paper mill operating until a time comes when it is economic to shut it down for a considerable span of time.

Under these circumstances, a low cost for wood from the forest is helpful for an integrated firm, but it is not critical. Low wood costs are desirable, as are low costs for any other input. But if wood costs rose 10 percent, or 25 percent, this would generally be far less serious than if wood supply fell, for some reason, by 10 percent or 25 percent. An available supply of wood in the volume dictated by efficient processing and marketing is vastly more important,within a wide range of price, than is the lowest possible price for that wood.

An integrated forest industry firm looks very much to the markets for its output. For lumber and plywood, whose prices are notoriously unstable primarily because of the highly volatile construction industry, the volumes that can be sold and the prices that can be obtained are often positively correlated in the short run. That is, when the construction industry is booming, the volume of lumber and plywood that can be sold is high as is the price at which they can be sold; when the construction industry is in the doldrums, both volume and price are down. The volume and the price of paper are much more stable, although both of these vary. Much paper is sold under long term contracts; when demand rises, the buyer under such contracts benefits considerably, both in terms of an assured supply and in terms of favorable prices. Paper consumers who depend on short term purchases in the open market may benefit when paper prices are low but may be badly squeezed if paper is unavailable or very high in price; this happened to some users of newsprint only a few years ago. The paper mill benefits from a longterm sale contract in the assurance of an outlet and of a price, although it too may be pinched if costs rise while the price of the paper it sells stays the same. Many vertically integrated forest industry firms manufacture various wood specialty products--special paper containers, as milk cartons, or windows and window frames, or doors and door frames, and others. Especially for these specialty products, the cost of the wood that enters them is very low in comparison with the great amount of labor and capital that is embodied in them.

Forest industry products are not manufactured for storage to the degree that agricultural products are placed in storage or even that metal products may be produced for stockpile. That is, when the demand for lumber is down, the vertically integrated forest industry firm does not long continue to produce for inventory, but instead cuts its rate of production. A paper mill may produce a little longer for inventory, when sales are slack, rather than cut back its operations. But inventory or storage are not used as a major means of ironing out the discrepancy between a fluctuating demand and an economic rate of continuous production.

With major economic advantages to the timing of timber harvest for processing and marketing economies, and with no sharp timing economies in the growing/harvest cycle (see figure 1), the vertically integrated forest firms choose their harvest dates, within very wide limits, for processing and marketing efficiency. The economic date of timber harvest, as based on the age-yield curve and in consideration of the costs of timber growing, is important, even significant, but it cannot be dominant in decisions about date of harvest of industry-owned timber. The firm will cut more or less of its own timber, or buy more or less timber from others, for market and processing reasons, not for efficiency of growth in the timber stand.

In this general situation, there is a special case of the paper company which grows about half of its wood and buys about half--a common situation, especially in the South. Such a firm may have a local monopsony position, since it may be the only buyer of timber within an economically rational transportation radius; or it may have a local oligopsony position, in being one of only two or three firms buying timber from the same supply area. In either case, the firm or firms must offer growers a price for wood high enough to keep them producing wood over a period of time. But in each situation there are substantial advantages to the firm, and some advantages to the growers, in stable prices. As noted earlier, pulpwood prices have, in fact, been rather stable over time, rising gradually but steadily. When the demand for pulpwood changes, the integrated firm may decide to use its own timber in larger volume, if demand has risen, or to use less of it, if demand has fallen, while keeping both volume and price of purchases of wood fairly stable. It may well find it more profitable to use some of its own timber, that otherwise could be used as sawtimber and apparently produce a higher value outturn per log, rather than to raise the price to producers. For instance, suppose the firm had been buying 15 units of wood at a price of $10 per unit, and producing 15 units of wood also at a cost of $10 per unit. If demand rises by 10 percent, the integrated firm might divert to its papermill wood that would otherwise go to its sawmill, if that wood had a value up to 20 percent or even 30 percent more per unit than the pulpwood, if by so doing it avoids raising the prices of the pulpwood it buys. It might equally decide to harvest some of its own timber at a somewhat economically irrational date, for the same objective of keeping prices of purchased wood stable.

An integrated forest industry firm producing lumber and plywood might also harvest some of its timber at an age either younger or older than the optimum economic age, if demand for its products, or supply of stumpage from other sources, or other circumstances so dictated. That is, a firm might postpone harvesting a particular timber stand for a few years beyond the apparent economic optimum age, if it had other timber even more ripe for harvesting which would fully meet its processing needs. Or, if the available supply of stumpage was less than desired, it could make good some part of this deficiency by harvesting of some of its own comparatively young trees. There are clearly limits, both physical and economic, to this adjustment of harvesting date to meet other objectives than economics of timber growing, but within those limits considerations other than the economic date of timber harvest might well be dominant in the firm's harvest decisions.

Willingness of a Vertically Integrated Forest Industry Firm to Invest in the Growing of More Timber

Why should a forest industry processing firm grow any timber? Why should it not buy all of the timber it wanted? If a processing firm could indeed buy all the timber it wanted, when it so desired, then presumably the only argument for growing timber, instead of buying it, would be that it could be grown cheaper than it could be bought.

Even under these circumstances, there might be some modest economic advantages to a vertically integrated forest industry firm in growing rather than in buying timber. Timber varies enormously in the United States, as to species, size of logs, grade of logs (including freedom from knots), amount of defect (rot, etc.) in the logs, and in other factors. It might be argued that a processing plant, knowing the character of the logs it would receive, could be operated to get a higher value outturn from them than could any other purchaser, less well informed about the logs in the woods. Or, alternatively, it might be argued that a processing plant could be designed and built which would get a higher value from any given lot of logs in the woods, than could any plant not so designed. In vertically integrated forms in many fields other than timber processing, there are indeed economies due to a close integration of raw material supply and processing plant. Some of these economies of vertical integration are achievable by log trading whereby each firm acquire logs of the type it can process most efficiently. But log trading also involves costs.

Another advantage to an integrated forest firm in owning at least part of its timber supply is that is would be less vulnerable to shortrun fluctuations in stumpage prices. Over a long period, the logs the firm grew should be priced or valued for its internal decision-making at the average price for which similar logs could be bought from another supplier. But the kind of variations in stumpage prices shown in figure 3 are upsetting to log processors as well as to log producers. Some reserve supply at a cost which did not vary so much in such short periods of time would clearly be advantageous to any mill operator.

Far more important than these possible economies of vertical integration of supply and of processing plant, the fact is that forest land and timber are not readily available in desired quantities especially in the shortrun, to would-be purchasers in many situations in the United States.

The market for land is always imperfect, in the economist's sense of the term, and this is certainly true for forest land. A timber processor typically cannot buy all the timber he wants, of grades, species, and sizes of logs wanted, at all times. A processor wholly dependent upon purchase of timber would almost surely be in serious trouble for lack of an adequate supply, if not typically then surely at intervals during the life of the processing plant. Any firm able to make the substantial investment required for a vertically integrated forest product processing plant is not going to trust the continued operation of that plant to the availability of logs in the market, if it can operate its forests for at least part of its supply. In the larger vertically integrated forest processing firms, the value of the going concern in processing and marketing is likely to far exceed the cost of owning at least some of the timber growing capacity needed.

Even when a vertically integrated forest firm would like to own a considerable part of its timber growing requirement, circumstances might limit its capacity to do so. All the existent integrated forest processing firms have a history, of course. Some of the present firms acquired a lot of timber and of timber growing land long ago. This is true, for instance, of one of the largest such firms, the Weyerhaeuser Corporation, which bought extensive timber holdings in Washington shortly after the turn of the century and has added to them over the invervening years. Some other large firms have been build up, primarily by consolidation of existing firms, in relatively recent times, and none of the predecessors may have been large timber owners. The Boise-Cascade organization, for instance, is heavily dependent on purchases of national forest timber.

Even if such firms wish to acquire forest land and have the capital to do so, it may be impossible or nearly impossible for them to buy incremental areas of land, in desired locations and of productive capacities desired. These large vertically integrated firms may be able to buy up the whole of some other firm, and consolidate the two, when they could not buy any of the forest land and timber of the other firm.

Along the Pacific Coast until roughly 1950, there existed substantial acreages of privately owned timber, much owned in modestly small tracts (a very few thousand acres at the most, often a few hundred acres per tract). The initial methods of transferring such land from federal public domain to private ownership had typically been in tracts of less than 500 acres; even when railroads had received large grants for aid in railroad building, they had sometimes sold such land to private owners in similar size tracts. A timber processor often could buy all the timber he wanted, when he wanted it, and in approximately the location he wanted; and he was in no hurry to buy, letting the landwoner bear the holding costs of the timber. In the period from 1930 to 1950, more or less, there was far more profit to be made from timber processing than from timber growing. Suddenly, as the end of the privately owned old growth timber came in sight, about 1948 to 1950, there ensued a scramble to buy up such private forest land. Firms which were too slow to act, or too slow to realize what was happening, or lacked the capital for timber land purchase, woke up one day to find that the only timber they could buy was federal timber, unless they were prepared to buy out another firm in its entirety. A considerable excess capacity in timber processing, in relation to volume of federal timber sales, resulted, and many firms encountered a real squeeze.

As pulp and paper mills began to move into the South thirty or forty years ago, they were desirous of having at least some timber growing land under their full control in order to insure some supply of stumpage at all times. At the same time, for public relations or political reasons they were hesitant to buy enough forest land to produce their full supply of logs. Large scale purchases over short periods would almost surely have pushed forest land prices up. Often they could present an image of good neighbors, willing to help small local landowners, and at the same time get wood at very reasonable prices, by buying cordwood from growers. Some of them leased land from small owners, planted and managed the forests, and paid an annual rental to the owner in order to get wood. By having these different sources of supply, they were pretty well assured of an adequate supply of wood at all times, able to buy wood at reasonable prices, and acquire a substantial measure of local goodwill.

For a vertically integrated firm contemplating the growing of its own timber, the processing plant may become a variable. Shall the firm add a new plant or plants, to those it already has? If so, where? what type (for example, what size, grade, etc. of logs to be processed)? how large? How fast should such a firm seek to process the timber it now has and the timber it expects to grow? Holding timber for later harvest, or growing to older ages, entails costs, hence implying that early harvest would be economical. But early harvest would require a relatively large processing plant, whose full technical lifetime might not be utilized before the timber was all gone. Would a smaller plant, more fully utilized over its full lifetime, result in savings in processing which would offset, or more, the holding costs of the timber? As a vertically integrated firm contemplates buying forest land, investing for timber growing on land it owns, or investing in new or remodelled processing plant, these types of questions and of tradeoffs become very important. Obviously, the range of alternatives that one can imagine are very wide, and this is not the place to attempt a detailed consideration of even the more common such alternatives.

For decisions of this type, however, the economic analysis as outlined earlier is very helpful. Timber yield curves, for specific sites within the existent or proposed timber properties, would be essential. A large firm would very likely encounter highly variable production sites within its total property, and management plans would have to be based on such site qualities. The optimum management practices, including optimum intensity of timber production, would also have to be considered. Likewise, the optimum age for timber harvest would be an important management decision. For all of this, the economic approach would be highly valuable.

But even here, the vertically integrated firm might wisely decide to shorten or lengthen its timber rotation, at some cost in economy of timber growing, to achieve a greater degree of economy in timber processing. This would be particularly true for the first rotation from the mature forest (which might well be fifty years or more) but to some extent it might be true for later rotations as well. As we noted in the discussion of table 1 and figures 1 and 2, there often is a period of years during which the economics of harvesting at the best date are not much better than harvesting a few years earlier or later. The timber growing process, while important, is only part of the whole. The vertically integrated firm must look at the whole of its operation, from woods to final market for its products, and seek to maximize profits for the whole enterprise. This may, or may not, be the most economical method of growing timber, if the latter could be considered in isolation. The economies in timber-growing would have to be compared with the economies of timber processing.

Important Characteristics of National Forests Today

The third forest management situation considered in this report is the national forests. As pointed out previously, they include about 18 percent of the commercial forest land in the United States, and about half of the standing softwood sawtimber volume [9]. These national figures, while accurate, do not present the full picture of the importance of the national forests. There are really two national forest systems, one for the Eastern half of the United States, the other for the Western half. Some of the Eastern national forests were reserved from scraps of public domain, as in the South and Lake States, but most of their area in the East was purchased from private landowners. In the East, national forests are less than 5 percent of all commercial forests. Much of the land purchased had been denuded of harvestable timber, which was one reason its owners were glad to sell it; in spite of a program of building up the stands on these lands, in 1970 their stocking rate was only about average for all commercial forests in their regions.

In sharp contrast, in the Western half of the country national forests are more than half of all commercial forest area, reaching to two-thirds in the Rocky Mountains and to nearly the entire area in coastal Alaska. In the West, the national forests contain a great deal of mature old growth timber. While direct comparisons with figure 1 are not possible, a considerable area of the Western forests contains trees at stocking rates for their species and sites comparatively well to the right of area A. That is to say, they have stands which have passed the age of mean annual increment in growth, and some stands have reached or passed the point of maximum standing volume. Net growth is low, or zero, or even negative on many older stands. For the coastal Douglas fir region of the Pacific Northwest, nearly half of the gross annual growth is offset by mortality, so that net growth is not much above half of gross growth. By any standards, much of the timber on Western national forests is older than the silvicultural optimum age and still more in excess of the economic optimum harvest age.

With such a large fraction of the total forest land, the national forests would inevitably exhibit some special characteristics of management. In contrast, no other forest owner, public or private, has more than about 1 percent of the total commercial land acreage. By any economic standard, the national forests must be a dominant firm; are they also an oligopolist?

The Forest Service, as the managing agency for the national forests, has no processing facilities for its timber. The federal government has never sought to establish processing plants capable of processing all the timber produced on the federal lands. Instead, the timber is sold to private firms. A symbiotic relationship thus exists between the Forest Service and the private timber processors. The Forest Service is dependent upon them to buy the timber from the national forests; they are dependent in varying degrees upon the national forests for timber which they may convert into salable products. There is a complex history of these relationships and a complex pattern of them today, which it is not possible to examine in detail here.

The national forests operate under a body of law which requires that timber be offered "at competitive sale." For various reasons, a competitive sale in the economic sense of the term is impossible. There is but a single seller, the Forest Service; and frequently there is but a single buyer, or only a few buyers, although sometimes spirited bidding does take place. Logs are comparatively heavy in relation to their value, hence transport costs alone tend to limit markets to local areas. Collusion among possible buyers has been charged in several instances, and proven in a few; as Mead has pointed out, outright collusion need not exist, if potential buyers temper their bidding on one sale in the knowledge that their potential rivals may temper their bidding on another sale [10].

There is considerable confusion as to the role the Forest Service plays on national forest timber sales. It makes offerings of timber for sale; sales are made only when a buyer offers a price equal to or greater than the minimum appraised price. Sales thus depend upon buyers. In practice, offerings depend upon the agency's estimate of possible buyer interest, because there is little point in spending public funds to offer timber no buyer will buy, so that even offerings are partly governed by prospective buyers. Harvest of timber under sale contract lies partly in the decision-making province of the buyer; he can harvest quickly, or more slowly, within limits prescribed in the contract, dependent upon his estimate of market demand for the forest products. At any date, there is likely to be twice as much timber under sales contract from national forests as an average annual cut. The forest products industry has pressed for more volume under contract, to permit buyers more flexibility in timing of harvest.

In offering timber for sale, the Forest Service engages in an elaborate timber appraisal process, and the appraised price of the timber is normally included in the advertisement of the sale. If there is but a single potential bidder who is in an economic position to process the logs, he normally bids the appraised price. It does not matter whether the sale is by sealed bids or by oral auction, if he is sure he will be the only bidder. If the potential bidder is less certain that he will be the only bidder, then at sealed bid sale he may bid higher, thinking this is necessary to shut out suspected competition. If in fact there was only one bid, this bid will be above the appraised price, sometimes much above it. The records of federal timber sales show that sealed bids with only one bidder mostly run at appraised price but a few much higher, for this reason. If the sale had been by oral auction in this case, the sale would have gone for the appraised price, since no competition would have surfaced. In those cases when there is great demand for national forest timber and potential buyers are prepared to bid against one another, either sealed bids or oral auctions will produce prices well above the appraised price. The appraisal process starts with the expected value of the timber products that can be produced from a given lot of logs; transport, harvesting, and processing costs are deducted to arrive at a stumpage price in the woods. The costs include a full depreciation allowance for all the private capital involved in log transport and manufacture, as well as a normal profit allowance. A processor desperate for logs can rationally bid away some of this margin rather than lose the logs. There have been instances when the competitiveness of an oral auction led potential buyers into bidding prices for national forest timber (and other public timber) to heights that left them very little operating profit. The possibilities of sale are thus numerous, ranging from outright collusion to restrict competition at one extreme to cutthroat competition at the other extreme.

There is a great myth, sometimes in the Forest Service but even more outside of it, that national forest timber sales are made at competitive prices. In fact, when one supplier of any raw material owns half of the potential supply (and a vastly larger proportion of the supply offered up for sale, since integrated firms do not normally sell any of their timber), that single supplier makes the market to a large, often to an overwhelming, degree. As the President's Panel on Timber and the Environment said: "...the effect of sales policy for national forest timber upon the total timber supply and demand situation can no longer be ignored" [11]. No matter how competently and skillfully the Forest Service offers its timber for sale, and even when there is a reasonable approximation to active competition among buyers, the Forest Service has dominated the timber market, simply by the volume of timber offered for sale. Were its sales only half as large, prices for logs would be much higher, especially in the shortrun; were its sales twice as large, prices would be much lower, again especially in the shortrun. In any decision-making process for timber sales from national forests, the "market" cannot be taken as an independent of the Forest Service's decisions as to volumes sold. Instead, the effect of the agency's actions upon the total market must be considered in the decision-making process.

The Forest Service as a Far-sighted Oligopolist Since 1905

In 1905, responsibility for the administration of the forest reserves was transferred from the General Land Office in the Department of the Interior to the Forest Service in the Department of Agriculture, and two years later the reserves were renamed as the national forests [12]. At that latter date the reserves included about 150 million acres of land [13]. Considerable acreages have been added since then, partly through the operations of the Weeks Act since its passage in 1911, partly by addition of the grasslands purchased under the submarginal land purchase program of the New Deal, and partly through additional withdrawals of public domain. On the other hand, considerable acreages of land have been eliminated from the 1907 reserves, as experience demonstrated they included some land not suitable for forest or grazing use or included land suitable for homesteading or other private use. To a considerable extent, the reserves had taken form in 1907 and by 1910 their total acreage was about as large as it is today.

Suppose that in 1907, instead of the forest reserves being placed under new federal agency administration, they had been sold in one block to some private individual or firm, who would have been free to manage them as he chose. Such a private owner would clearly have been an oligopolist, at least in many areas of the West. How would he have managed the national forests differently than in fact the Forest Service has managed them, under applicable laws.

Had the forest reserves been offered for sale in one block in 1907, or even in several smaller blocks, it is very difficult indeed to guess what they might have been sold for. They were remote from urban centers, from timber manufacturing plants, and from established transportation routes, in most instances. There was an enormous volume of vastly more accessible private timber on the market, available for purchase by any processor who wanted it. Given the prices at which such timber was selling, and given the prices at which large tracts of public and private land had been sold (Texas, for instance, between 1870 and 1890, had sold a great deal of its state land, not forested but valuable for grazing and sometimes for farming, at prices well below $1 per acre), it is doubtful that as much as $200 million could possibly have been realized by such a block sale.

It is interesting to observe that in 1900 Frederick Weyerhaeuser and associates bought 600,000 acres of prime forest land (and its timber) from James J. Hill, the builder of the Northern Pacific Railroad, for $6 per acre and it continued to add to its holdings over the intervening years [14]. The land so purchased averaged much higher in timber volume, timber quality, accessibility to water transportation, and productive capacity to grow timber in the future than did the forest reserves as a whole. Half of the latter were not commercial forest at all, but grazing lands primarily. Large scale private purchase of timberland was comparatively unusual in the early years of the twentieth century; money was mostly to be made in processing timber and at that time there were many ways to acquire timber for processing without owning large acreages of land.

Suppose the forest reserves had been put up for sale in a block; who might have bid $200 million or any figure reasonably close thereto?

The big fortunes of modern times were then nonexistent; Rockefeller was just making his fortune in oil, probably had nothing like this amount of capital, and would have judged (rightly) that he could make far more money in oil than in lumber; Carnegie was emerging with a fortune from steel, but of nothing like this size; Henry Ford was a struggling mechanic; and Howard Hughes and Paul Getty were undreamed of. True, considerable capital had been accumulated from private sources for railroad construction. But timber would surely not have appealed to many people as a place for a financial killing.

So an image of sale of the forest reserves in 1907 in one block, to one buyer or to a few buyers, is economically nonviable, as well as politically unrealistic. But, never mind, let us pursue the question, what would this hypothetical new owner have done with his huge holding? One alternative might have been to cut the timber as fast as he could, retain some of the most productive lands for continued timber growing, and sell off or give away the rest on the best terms he could--the liquidation strategy. This is what might seem to result from an unthinking application of the traditional economic analysis; it is, moreover, what most timberland owners were doing about this time. Transform as much as possible of the capital tied up in trees into more liquid capital, and use it elsewhere where rates of return were higher. The trouble with this strategy is that it was impossible on large scale, given the acreage and volume of timber on the forest reserves, and given their generally remote locations. A liquidator could not have gotten very far at that time.

Our hypothetical oligopolist might have chosen a drastically different strategy. He might have said: sell a little timber each year, as demand permits, to meet the cash holding costs, or as nearly so as possible;

but retain the vast bulk of the timber for a distant day, when the vast private virgin timber holdings would be largely exhausted, and when his reserves would be highly valuable. This is the hoarder approach. To build a good public image, so as to keep setting of forest fires to a reasonable minimum and to help persuade legislators to go easy on taxes, the oligopolist might well have welcomed hunters and fishermen, made some modest provision for other recreationists, even set aside some of the completely unharvestable areas as wilderness, and talked a lot about multiple use. Following this course, the oligopolist would have had one trillion board feet of standing sawtimber in 1975, worth about $40 billion for land and timber [15]. It is extremely difficult to say what interest this hypothetical oligopolist would have earned on his original investment; through much of this period, especially the early years, cash deficits from operations would have been nearly inevitable. While small, they would have added to his investment. By the mid-1940s, he might have been making a cash surplus annually from forest management (the Forest Service did this by 1952) and with frugality he might have continued to achieve a cash operating surplus annually right up to 1975 (the Forest Service surplus disappeared by 1961 but perhaps a private owner could have done better [16]. Ignoring cash deficits and surpluses, or assuming that they cancelled out, the oligopolist would have earned, in current dollars, something in the rough order of 8 percent interest compounded annually on his original investment--not sensationally high, especially considering that his 1975 dollars were worth a great deal less than his 1967 dollars, yet not so bad for a seventy year period.

For reasons outlined above, this picture of the far-sighted oligopolist, hoarding the timber on his reserves against a distant day when the timber will be more valuable, is only a myth, a dream, an image, wholly impossible of achievement, even if it had been desired. But its similarity to what has actually been done on the national forests, by the Forest Service operating under relevant laws and applying the standards of "good forestry" as developed by the forestry profession, is very close indeed. The reasons for the public agency doing what it did were different from the reasons the far-sighted oligopolist hoarder would have used, but the results over the years and the resulting present situation are very closely similar.

Applicability of Biologic and Economic Analyses to National Forest Management

Close application of either a biologic approach, such as maximization of MAI or of an economic approach, such as determination of financial maturity, to the national forests at any time during the past 75 years, would have been very difficult for either the Forest Service or for a private oligopolist, both for decisions on how rapidly to harvest standing timber and for decisions on how much to invest in growing more timber. The national forests 75 years ago, and continuously to the present, presented a forest management situation not really contemplated by either the biological or the economic models of timber harvest.

The western national forests had and have an immense volume of standing timber which has been inherited from the past--in excess of 950 billion board feet in 1970 (the latest year for inventory data) or more than half of all the standing softwood sawtimber in the entire country. This timber is slow growing, with rot and other mortality taking nearly half of the small gross growth which does occur. The usual inventory statistics, while accurate, in some degree underestimate the nature of this very high inventory situation. Some of these national forests have been cut during the past fifty years or so, and, regardless of the success of forest regeneration, these sites do not now have the typical very heavy stands of mature old growth timber. For the national forest areas which have been uncut (and which have escaped recent fire, insect, or disease destruction of standing volume), the volume-age relationship is far to the right of area A in figure 1; it falls in or close to Hishleifer's second order of absurdity area, where annual growth is very low and sometimes negative [3].

Given such age-volume-growth relationships on a forest, the usual biologic or economic prescription for a small private forest would be unequivocal: cut and sell promptly! But this has been and is impossible for the national forests as a whole, even if there were no laws forbidding it. Their timber volume is so large that processing capacity is inadequate to handle it all within a short time, and the effect of putting so much timber on the market in a short time would be a catastrophic fall in stumpage prices. Even trying to sell that land and timber together on a large scale quickly would depress prices. The real questions arise with how to manage this old growth timber and more explicitly with how much of it should be cut and how rapidly. In a later section, we analyze the future possibilities under the National Forest Management Act of 1976.

Before considering what the Forest Service should now do with this volume of timber, let us go back to our hypothetical oligopolist, and assume in 1977 he still owns the national forests, with their present stand of timber. Today he would almost surely decide to liquidate this stand of timber, in spite of current projections that the price of stumpage will rise in the future. Indeed, he probably would have come to this conclusion long before 1977, and this in spite of the fact that at almost all dates since 1907 the projections were for stumpage prices to rise. Without facing the question of when in the past he would have decided in favor of a liquidation policy, we can be fairly sure that in 1977 he would have concluded that prospective future rises in stumpage value would not offset the very low physical and economic yield of these forests. Over a period of time, he would undoubtedly sell every mature tree that could be harvested, transported, and processed at a positive return. He might well disregard all values in his forest other than timber, on the rationale that, if there was truly some social value to the retention of some part of his stands, the public should either pay him for such values or should restrain him by law from harvest where there was indeed a clear social value demonstrable.

But even this hypotehtical oligopolist, bent on early liquidation of his timber stands and their conversion into capital investable at a higher return elsewhere, would have to consider at least two other major matters:

1. What is the demand for the timber from his lands? How much would his accelerated timber harvest depress the price of wood and its products [6]? The answer would depend in part upon the reaction of other forest owners, upon their willingness to harvest their existing stands of timber and their willingness to invest in growing more timber. If the demand for logs for the next ten to twenty years seemed to this oligopolist likely to be moderately elastic, and if the willingness of other forest owners to harvest their timber was also moderately elastic (so that, when prices of stumpage fell somewhat, they would cut back their harvest more than proportionately), then he could move a far larger volume of logs into the market profitable than would be the case were these elasticities much lower. In any case, he would have to play off

declines in stumpage price against immediate gain from harvest which would permit investment of the funds so received at higher rates elsewhere. Even if the hypothetical oligopolist tried to liquidate the western national forest timber as fast as was economically rational, he could hardly do so in less than twenty years. Even if his objective was only to liquidate slow growing over-mature timber, during any reasonable liquidation period he would be faced with considerable further growth of timber. A twenty year liquidation period would mean an average annual sale of stumpage from national forests of 50 billion board feet or more--three to four times anything ever experienced to date. If he decided to sell both land and timber to another private owner, the latter would face many of the same problems.

2. Even though our hypothetical oligopolist had no timber processing facilities of his own, he would have to consider the economics of timber processing, for the people to whom he sold the stumpage would have to process it. What bait would he have to hold out, to possible buyers, to induce them to make the necessary investments to build a processing capacity adequate to handle the volume of wood he chose to sell? Such bait might have to be more than price; it might well have to include some commitment or contract as to volume and price of timber that would be offered for sale annually.

The Forest Service has never been free and is not now free, even if it so wished, to follow a liquidation policy for the timber on the western national forests. The Multiple Use and Sustained Yield Act of 1960, the Renewable Resources Planning Act of 1974, and the National Forest Management Act of 1976, to name but three acts, require the Forest Service to manage the national forests for continued future production of timber, as well as of other products. The political support for the national forests is based upon a continued productivity of these lands and upon a continued supply of products from them. Lastly, and by no means least, the Forest Service as an organization is philosophically and ideologically committed to continued timber production. So liquidation of existing timber stands a la oligopolist, is out; but this does not mean that maintenance of present volumes of timber on the western national forests is necessary, either to comply with the laws or as part of a sustained yield philosophy.

Before either a biologic or an economic rationale can be applied to the matter of how much timber to cut annually from the western national forests, national policy as expressed in these acts requires that consideration be given to at least the following factors:

a. Areas of national forest where no timber harvest at all should take place, because of conservation factors. That is, lands too steep or with unstable soils should not be harvested at all, because the resultant erosion would be unacceptable. Or harvest should not take place too near water bodies such as lakes or streams, because the adverse effects upon water quality would be unacceptable. Definition of such areas would, of course, pose problems and likely to lead to some arguments, even if the general idea is accepted. Similar constraints on timber harvest on private land are gradually being applied by public actions.

b. Areas of national forest where no timber harvest at all should take place, because the value of the standing timber for uses other than timber production is greater than the values that can be realized by timber harvest. On some areas outdoor recreation, wilderness, and other uses conceivably can produce higher annual values from the presently standing timber than can be realized from its harvest. There would also be value in maintaining some stands of timber, simply as relics of a past kind of forest. Forest areas, in both public and private ownership, have been set aside in many parts of the country for their purposes. Again, even if there were agreement on the general ideas, there might well be considerable disagreement on specific areas.

c. Where timber regeneration, naturally or from planting, is too slow or too uncertain, one should at least be cautious about timber harvest; and unless the values of the standing timber are very great, they might well be foregone. No small part of the public opposition to clearcutting has arisen from situations where regeneration was long delayed.

d. Lastly, on those areas of national forests where the future costs of growing and managing timber for timber harvest are simply too great in relation to the values produced thereby, such areas might well be set aside and reserved from cutting of the present stand, at least for the present. The Forest Service is committed by law and by its own philosophy to the growing of timber indefinitely into the future everywhere on the national forests where trees grow "naturally". The present harvest of timber on poor sites might yield a (probably small) net return, but it would constitute a commitment to larger future expenditures than would be recovered by later timber growth. The present harvest is in effect a future promissory note. The available data do not permit a percise estimate of this latter kind of area; much, if not all, of the Site Class V national forest (31 percent of the Western national forests) would probably fall in this category; while some site class V land might grow timber economically, some site class IV probably could not do so. More exact determinations would have to be made, site by site. The quality of the timber grown, the cost of getting it out, and other factors would be involved.

The combined effect of these four reasons for not harvesting exist ing stands of Western national forest timber cannot be estimated closely; for one thing, there is an unknown degree of overlap among these categories; Some present forests might be eliminated from harvest for all four of these factors--too steep, too valuable for other purposes, too uncertain for regeneration, too uneconomic for continued timber growth. In other cases, only one reason might be involved. For instance, there might be a national decision to retain some old growth timber for wilderness even though continued wood production would be economically profitable and acceptable conservationally. But, even when full allowance is made for all of these factors, there remains a substantial area and a large volume of Western national forest timber which can be harvested and where timber can be grown for harvest in the future, economically.

On the Western national forests, land in site classes I through III include 39 percent of the area of all commercial forests; because their productivity per acre is so much above that of the land in site classes IV and V, these more productive forests include 67 percent of the productive capacity of all forests under natural stand management and almost surely a larger proportion of the productive capacity under intensive management. While data are unavailable, it seems probable that the better class lands contain as much as 60 percent of the truly harvestable and merchantable timber on all the Western forests. The best class sites not only have heavier stands per acre than the poor sites have, but the timber is likely to be of higher quality, with less rot and reject material. As a result, such timber has a higher value per unit of quantity. Let us assume that all site class IV and V Western national forest land is unharvestable for one or more of the reasons outlined above, or that any harvestable volume on such lands is offset by some unharvestable volume on the more productive sites. This is an assumption which most of the forest industry would consider as excessive reservation of timber from harvest. I intended it to be on the high side. But it still leaves about 570 billion board feet of presently standing softwood sawtimber available for early harvest on the Western national forests.

If the harvest schedule for this accumulated volume of old growth timber were to be accelerated, the Forest Service would have to consider the same factors as outlined for our hypothetical oligopolist--demand for timber, and processing capacity and processing economies. In planning year to year variations in harvest volume, the Forest Service might well consider probable year to year variations in demand, as well as longer term demands.

Past Management of the National Forests by the Forest Service

Faced with an inherited timber situation on the national forests which did not remotely conform to the "fully regulated" forest of the forester, the Forest Service formulated some new concepts or applied and developed concepts formulated by others, for its management of the national forests [17]. These concepts were applied both to decisions about harvesting existing stands of timber and to investment for the growing of additional volumes of timber. These concepts include sustained yield, allowable cut, even flow, and allowable cut effect, each of which must be examined briefly.

In its broadest and most general terms, sustained yield simply means a volume of timber harvest which can be sustained indefinitely from a given forested area. As such, it has great intuitive or emotional appeal, especially to nonforesters. In fact, there exists no such thing as the sustained yield, any more than there exists the yield of corn; production of both timber and corn involve inputs of labor, capital, and management on an area of land, and both sustained forest yield and corn yield are variable within considerable limits,depending upon the inputs. There are many sustained yields, not simply one. In practice, many foresters mean by "sustained yield" that yield which comes from reasonably good "natural forestry"--for example, keep the fires out, as far as reasonably possible, let the trees grow naturally as they will, harvest in ways which facilitate natural reproduction, and the like. But economically ally rational sustained yield may well vary from zero to double or more "natural forestry" yield. In practice, in its use of the term sustained yield and in its management of the national forests to attain it, the Forest Service has meant reasonably good natural forestry.

In its broadest and most general terms, allowable cut is the amount of timber which can be cut annually. Allowable cut is normally constrained by estimated sustained yield but it might be constrained by other or by additional factors. The Forest Service application of the idea to the national forests is fully presented in the Forest Service Manual [18]. The term applies to both the longterm planned harvest and to the programmed annual harvest [7]. The Forest Service generally requires that the trees reach some minimum size before harvest and bases its allowable cut on a planned rotation age for the particular timber stand. Since old growth stands are generally older than the planned rotation age, and since timber volumes average much higher than they will be under a planned rotation, a conversion period from old growth to rotation is necessarily involved. Numerous formulas, some quite complex, have been developed for calculating allowable cut under these circumstances. The conversion period on national forests is one and one half times the number of years to reach the culmination of MAI. This results in a slow harvest of the mature old growth timber, often over a period of 150 years.

Forest Service estimates of sustained yield and of allowable cut on national forests are based directly upon the kind of age-growth-volume biological data in table 1 and in figures 1 and 2, but they implicitly include some economic factors as well. The output of all kinds of products and services from the forested parts of the national forests has risen over the years [19]. Timber harvest has risen from around 1 to 1½ billion board feet annually in the 1920s to an average of over 11 billion board feet in the period centering on 1970. This increased harvest of national forest timber reflects the increased demand for it, as more accessible stands of private timber were harvested. Additional areas were opened for harvest by the building of roads to more remote areas, and some species of trees came to be utilized commercially when at an earlier date they were not considered commercial species. Over the years the Forest Service revised its estimates of sustained yield capacity of national forests, from an estimate of about 9 billion board feet annually in the mid-1950s to one nearer 15 billion board feet in the late 1960s [20].

Management of the national forests to maintain sustained yield capacity was clearly mandated by the Multiple Use-Sustained Yield Act of 1960.

The Forest Service developed the concept of even flow as a special version of the sustained yield and allowable cut ideas. In its most extreme form, even flow requires an exactly equal cut of timber each year. In practice, small deviations from year to year are allowed if the average cut over a period of years (usually ten) averages at or close to the allowable cut. Variations from year to year might be in response to the need for salvage, if salvage cuts are included in the allowable cut calculations, or in response to market conditions. The Forest Service seems to have believed that even flow of timber was required by the Multiple Use-Sustained Yield Act of 1960, but others have doubted that such a management program was really required by that act [11]. In practice, strict application of the even flow concept on a national forest by national forest basis results in substantially less allowable cut than does some relaxation of even flow either through time or by area of application.

The typically long transition period between the inherited mature forest and the fully regulated forest involves a heavy physical waste of timber which often disturbs foresters (as well as high capital costs, which seems to concern them far less). Some foresters, concerned with this problem, have invented the concept of "allowable cut effect" (ACE)[21]. The basic idea is that some practice, such as the planting of seedlings on a tract otherwise not regenerating (or even, presumably, on a tract never having borne trees in the past), is begun now, its effect upon sustained yield at some future date is calculated, and this amount is immediately added to the allowable cut of the inherited old growth timber.

Obviously, this would be impossible were there not a large volume of old growth available for cutting; other conditions for application of the allowable cut effect are specified by these authors. "ACE is the immediate increase in today's allowable cut which is attributable to expected future increases in yields....ACE is not limited to instances where bare land is brought into production. It can result from any management practice that increases average growth rates" [21]. By comparing the present cost of the management practice for the anticipated future harvest with the present increased income from the harvest of existent old growth, some highly favorable "benefit/cost" ratios can be attained.

As typically applied, the "allowable cut effect" does not consider the later costs of growing to maturity the timber which is expected ultimately to result from the present practice. That is, the cost of planting an acre now is compared with the value of the old growth timber that can be harvested now from an old growth stand. If the full costs, including the capital costs, of growing the timber to maturity were included, of course the benefit/cost ratio for planting would be very much less; the economic result would be the usual one of analyzing benefits and costs over the rotation period of the timber stand.

The allowable cut effect approach to timber management is fallacious. It is simply a rationale for cutting inherited mature and over-mature timber faster than the rate which the sustained-yield-allowable-cut-even-flow approach has indicated. Foresters, having adopted a management philosophy and plan which results in very long periods of cutting the inherited mature stands, with the attendant high costs and wastes, now seek some rationalization of abandoning their self-imposed limitations. Speeding up the harvest of inherited old growth timber because of "allowable cut effect" accomplishes nothing that the same acceleration of the old growth harvest would not have accomplished in the absence of the allowable cut effect measure.

There are many situations in life where some group or individual, having adopted some course of action, then seeks a rationale for abandoning it. For instance, within the past year or two, someone has seriously proposed compulsory installation on private autos of a device which would limit the car's motor from developing more power than would carry the car along at a top speed of 55 miles per hour. The proponents of this idea have argued that such devices could be made and installed for a cost per car of about $25.00, and that, at this price, such devices would be "highly cost effective". No doubt this latter statement is true, if one grants all the premises. Having built a needlessly powerful car, at the cost of some hundreds of dollars more than would have been incurred in building a less powerful one, we now propose to hobble the car with an additional device, to bring it to a speed which we regard as desirable. The proponents of this idea seem not to have faced the possibility that the same political muscle and the same popular acceptance (or non-acceptance) problems would be encountered in the application of their idea as would be involved in compulsory restriction of auto design and construction to achieve the ends they sought, and that the latter might be vastly more "cost effective" than the control device. If we really wish to accelerate the harvest of inherited old growth timber stands, to avoid the heavy physical and economic costs that their slow harvest entails, why invent "allowable cut effect" to persuade us to do so?

Those who apply the allowable cut effect do not, typically, consider shortrun markets or timber processing economies, any more than do those who calculate sustained yield and allowable cut.

The allowable cut effect calculations should never be allowed to justify forest management practices which will not stand on their own, as economically sound over the full lifetime of the stand. This includes proposals to plant trees on sites where natural regeneration is not producing a stand, on the grounds that the "allowable cut effect" makes this practice feasible economically. A forest manager should not plant sites on which the cost of growing timber to harvest, including interest on realizable capital during the growing period, exceed the expected values of the timber at harvest time. Leaning on "allowable cut effect" does not repeal the effects of time and the interest rate.

In its management of the national forests the Forest Service has always placed a great stress on "community stability". In part, this emphasis by early Forest Service leaders grew out of their rejection of the "cut out and get out" system of private harvesting of old growth timber stands which had prevailed over so much of the United States. The communities built up during the harvest period were usually abandoned--and the land often laid waste, ravaged by uncontrolled fires. Community stability as applied by the Forest Service contains an explicit or implicit equation of the constant volume of timber flowing out of sustained yield with a constant employment and hence a constant income to the community. There are several obvious shortcomings to this approach: (1) a constant volume of timber may not mean a constant level of employment, if technology changes; (2) with the future (especially the future 150 years from now, implicite in long timber rotations) remaining so unknowable, it may well be that the present communities will prove not worth saving--this has happened in the past, with forest communities; (3) community stability could not in any case be guaranteed with national forest timber alone, since in nearly all areas it is but part of the total supply; and (4) since the rate of harvest of privately owned timber has been variable, a constant rate of harvest of federal timber may actually be destabilizing.

Sustained yield, allowable cut, and even flow have all be defended as contributing to community stability. A slow rate of harvest of the mature old growth timber has been defended as a means of avoiding or of minimizing a "drop off" when the old growth timber has all been cut.

The management philosophy applied to the national forests in the past has not been the only philosophy that could have been applied to them, even with the same basic legislation governing their management. In particular, more intensive forestry could have been practiced on the national forests, with significantly different results in terms of timber harvested under a sustained yield-allowable cut philosophy--and even under an even flow philosophy, untenable as that idea often is. The drop-off in timber harvest when the old growth timber has all been harvested is a function of the management program followed, not an inherent necessity of forestry.

The data in table 1 and in figure 1, are sufficient to permit some illustrative different forest management plans for the Douglas fir site class II forest show in that table (table 2):

1. Management Plan A is a very long rotation (164 years of growing period, 171 years between harvests, the seven years difference being required for natural stand regeneration) of good "natural" forestry. At the time of harvest, the stand would contain about 21,000 cubic feet of harvestable wood. The average annual growth of harvestable wood for

Table 2. Alternative Timber Management Possibilities Output, Site Class II, Douglas Fir

					Management Alternative	
					Intensive	
Item	A	B	C	D	High Estimate	Low Estimate
Stand growing roation--yrs.	164	50	127	60	50	50
Harvest to harvest rotation--yrs.	171	57	134	67	51	51
Annual volume of merchantable timber						
Final cuts--cu.ft. per A	123	155	146	165	333	275
Thinnings--cu.ft. per A	0	0	0	0	265	206
TOTAL	123	155	146	165	598	480
Total volume at each final cut, plus thinnings--cu.ft. per A	21,000	8,850	19,500	11,050	30,500	24,500

Source: Data from Table 1.

the whole harvest-rotation period is 123 cubic feet per acre. This is somewhat the kind of management that would exist on a national forest where the inherited stand of mature timber was cut over a long period of time and where the next rotation of "natural" forestry strove to produce about the same kind and size of trees.

2. Management Plan B is also "natural" forestry, also with a seven year period of natural stand regeneration after harvest, but with a relatively very short rotation (50 years growing period, 57 years between harvests). This system provides three cycles of growth and harvest in the same period of years as one cycle under management plan A. Due to the fact that the trees would be cut nearer their age of maximum mean annual growth, the average annual growth per acre would increase to 155 cubic feet per acre, or about 25 percent higher than under management plan A; the figure would be still higher, were it not for the relatively long regeneration period. Over a 171 year period, plan B would yield about one fourth more timber than would plan A, from the same area of land. However, the first harvest under plan B would be only 42 percent as high as the first harvest under plan A. That is, given a situation in which the forest owners/managers inherited old mature stands of timber, their first harvest would be a large one, while the next harvest under the conditions of plan B would be much smaller. This is the "dropoff" of which so much is made in some of the literature about national forest management.

3. Management Plan C is a "natural" forestry plan with a fairly long rotation (127 years growing period, 134 years between harvests) but considerably shorter than plan A. The initial cut is almost as high as in plan A and the average annual growth rate (between harvests) is

nearly 20 percent higher. Plan C shows the advantage of a somewhat shorter next rotation, each as compared with plan A, but otherwise the two plans are similar.

4. Management Plan D is two cycles of growth and harvest for one period in plan C. The same "natural" forestry is used. Plan D has both a somewhat higher initial cut and a somewhat higher average annual growth rate than plan B; otherwise, the two plans are generally similar, and what was said about B largely applies to D. Plan D has a higher average growth rate than B largely because there is less "down time", or time between harvest and stand regeneration, because fewer harvests occur during a very long period. On the basis of length of growing period alone, one would expect plan B to do as well, or perhaps slightly better, than plan D.

5. Intensive Management of forests is entirely different than the "natural" forestry common to all the foregoing management plans. The components and the possibilities of intensive forestry have been explored in a recent report of the National Research Council--National Academy of Science [22]. Hyde makes generally similar calculations, but also includes a great deal of cost and yield data of a site-specific kind [7]. Intensive forest management consists of a package of measures, each of which tends to reinforce and make more productive the other measures in the package. For a Douglas fir forest on site class II land, the following measures would be included in such a package, and their effects would be about as follows:

a. By planting young seedlings, instead of waiting for natural regeneration of the stand to take place, the regeneration period could be cut from the seven years assumed for the "natural" forestry, to one year, with a consequent gain in average annual wood growth over the rotation of 4 percent for each year saved in regeneration, or 24 percent in total for the whole rotation. Hyde points out that, since seedlings are normally two years old when planted, the regeneration lag is a minus one year, compared with natural seeding [7].

b. By planting seedlings at optimum spacing for the site, the rate of annual growth could be 20 to 30 percent more than for the kind of spacing likely to result from natural regeneration--too close in some areas, too sparse in others;

c. By fertilization at appropriate intervals, the average annual growth rate for the rotation could be increased by 15 to 20 percent, or more.

c. By thinning the stand at appropriate intervals, substantial volumes of usable wood could be harvested, that otherwise would be lost by natural thinning processes (death and decay of trees that lost in the competitive struggle), and the size of the trees remaining for the final harvest of the stand would be much increased over the size of trees at the same age under "natural" forestry. Hyde estimates that trees remaining at age 40, after a similar regime, would have average diameters equal to those of a natural forestry stand at age 75 years [7]. The average annual growth rate of usable wood for the period of rotation could be increased 30 to 35 percent by this practice alone.

e. Genetic improvement could in the longrun increase wood growth considerably, as improved new strains were developed. But even tree selection and reproduction of seeds from the best trees, in nurseries from which seedlings could be planted, could increase average annual wood growth by 10 to 15 percent during the first rotation.

The foregoing numbers come from the National Academy of Science report; they were stated there as conservative. Each is for the described practice alone; the combined effect of all these practices (the synergistic effect) would be considerably greater. These practices, taken singly, sum up to a doubling in rate of wood growth over the rotation. My own estimate is much higher. By a combination of these measures, I estimate that the following volumes of wood could be harvested at the specified dates in the rotation:

	High Estimate	Low Estimate
	(cubic feet per acre)	
20 year thinning	1,500	1,000
25 year thinning	2,000	1,500
30 year thinning	2,500	2,000
35 year thinning	2,500	2,000
40 year thinning	2,500	2,000
45 year thinning	2,500	2,000
Final cut	17,000	14,000
Total	30,500	24,500

The result of intensive forestry in this illustrative case is not only a greatly increased annual rate of harvestable wood growth per acre but also a large volume of timber remaining for the final cut in the rotation. The trees taken out in the thinnings will be comparatively small, but will produce 2" X 4" studs and chips for paper, and the trees remaining at the final cut in the rotation will be large. Their total volume at the final harvest will be less than the total volume of the final harvest under plans A and C, but definitely larger than under plans B and D.

These illustrative calculations will undoubtedly seem overly optimistic to some people. They apply to one forest type, on one site class, in one location. The kind of intensive forestry underlying them is not yet common and hence not yet fully tested under a variety of operating circumstances. It does call for a much higher level of management--for a higher level of managerial competence--and for larger annual inputs than the usual natural stand forestry. On the other hand, the amount of capital tied up in timber stand inventory is less under intensive forestry than under management plans A and C with their very long rotations and consequent large inventories. Particularly important is the coordination or synchronization of management actions to achieve the potential result. In some ways, however, these calculations of the difference between natural stand and intensive forestry under-estimate the difference. Figure 1 was based on a fully stocked natural stand, in actuality, many natural stands are not fully stocked, but have regenerated at something less, sometimes a good deal less, than full stocking. To the extent this happens, the growth rates of figure 1 are higher than typical natural stand forestry would produce. Intensive forestry, to the extent it achieves its potential, would thus do better than natural stand forestry by a larger amount than the comparisons in table 2.

The data in table 2 show that the much-discussed "dropoff" is wholly a product of the management system employed, whether on national forests or elsewhere. If the inherited mature timber stands are harvested and the area then managed on shorter rotations by about the same kind of "natural" forestry which had produced the mature stand in the first place then a "dropoff" surely occurs, even though average annual rate of wood growth is greater. But if the area, after the initial harvest, is managed intensively, no dropoff occurs--timber harvest at the end of 50 years is as great or greater in total volume as it was during the period of harvest of the mature stand. If an inherited stand were converted to intensive forestry over a 50 year period, there would be no dropoff.

The foregoing point may be illustrated by two very crude and simple even flow models for the national forests as a whole, one based upon natural stand forest management and fairly long rotations, the other based on intensive forestry and much shorter rotations (figure 4). The first means an average annual harvest of national forest timber of about 15 billion board feet annually, the latter means an average annual harvest of about 30 billion board feet annually. Each is built to insure no "dropoff"; that is, each contemplates a level of timber harvest, beginning immediately which can be continued indefinitely into the future. Neither makes as much allowance for conservation, wilderness, and other nonharvests on possible timber lands as probably would be socially desirable. Neither has been subjected to economic analysis; that is, each includes harvest from some lands on which the costs of timber growing probably exceed the income that can be produced by such timber growing. While the accelerated-harvest-intensive-management model has not been subjected to economic analysis, it does contemplate tree harvest at something nearer the economic maturity of the trees that does the even-flow-low-intensity model. Silviculturalists will no doubt regard the model of intensive management as crude and will find some flaws in it, but more precise and more professionally respectable models could be built, with the same general purpose and the same general result. The basic conclusion is that even flow, though undesirable in many respects,

Figure 4. Production possibilities, sawtimber from National Forests.

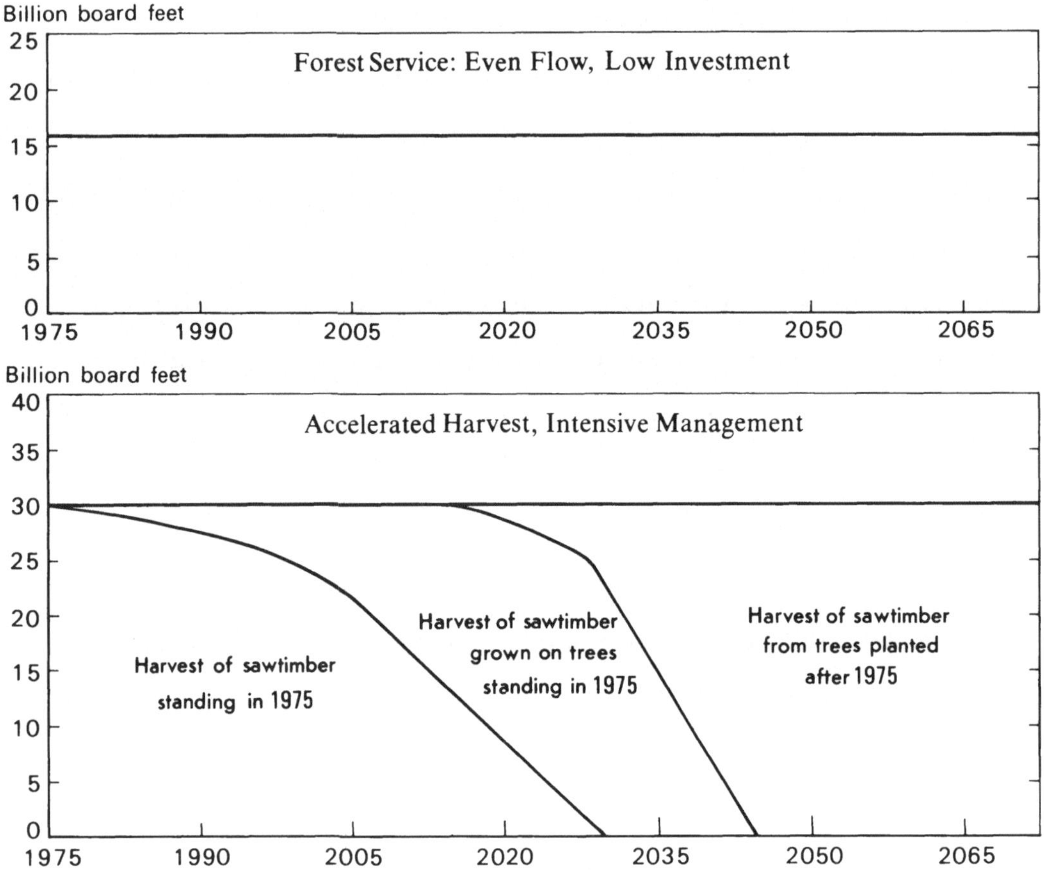

need not result in the kind of low level timber harvest that in practice it has resulted in. Within any reasonable economic and silvicultural constraints, a high output even flow model can be built.

The data in table 2 and in figure 4 do not include any analysis of economic efficiency; they simply present biologically possible outputs, without connotation of economic profitability. A full analysis of the economics of intensive forest management would be more detailed than can be included here, and in any event much of the needed data is unavailable. However, some calculations have been made which throw some light on the economics of forestry more intensive than "natural forestry".

Marty has calculated the percentage return on different kinds of forest management programs, for different sites and under different price assumptions [5]. His analysis is not confined to national forests or to the West, but some of the species for which he presents data--Douglas fir, ponderosa pine, and lodgepole pine, for instance--are western species found on large areas of western national forests. His results may be briefly summarized as follows: (1) rates of return are higher when wood prices are higher than when they are lower, and are very much higher when prices of wood are rising (as historically they have been) than when they are constant (even if high); (2) the effect of site quality is, generally, far greater than reasonable differences in wood prices--rates of return generally are twice or more higher on site class I than on site class III land, and he did not think it worthwhile to analyze returns for sites lower than III; and (3) rates of return are much higher for a number of silvicultural measures taken together than for any of them separately. Marty gets annual returns of 10 percent and more for a good many of the site, price, and management practices combinations which he analyzed.

Bill Hyde and I have estimated the future cost of growing wood in the Pacific Northwest and in the South, especially in the pine areas of the latter [23]. Our analyses were not directed specifically to the national forests but apply to them. In the retrospect of the few years that have elapsed since our studies were made, I wonder if we made adequate allowance for the economic effect of some of the newer forest management techniques; if we did not, the true results would be more favorable than our estimates. In general, we concluded that intensive forestry would pay on site classes I and II, but perhaps not on site class III and surely not on less productive sites in the Douglas fir areas of the Pacific Northwest and that it might pay on site classes I to III, but not on IV, in the South. Our studies in effect used only one estimate of future wood prices and only one set of cost estimates, and to this degree were not as detailed or as specific as we would have liked, had more detailed data been available. In his more detailed and precise later analysis, Hyde comes to a similar conclusion for the Douglas fir region [7].

Henry Vaux had previously made similar estimates of production costs and future wood production possibilities in California and had arrived at similar conclusions [24]. His analysis was not limited to national forests but included forest types common on the national forests of that state.

These studies and discussions with some of the more forward-looking timber industry firms lead me to the conclusion that intensive forestry on most of the site class I through III land on the western national forests will produce wood at costs which range from low to reasonable in comparison with probable future wood prices in those regions. At this stage of our collective knowledge, such a conclusion is about as far as one can go; it is undesirable general, but I think it is not the less accurate as a result. The specifics would vary, from site to site and from one area or region to another, and the general statement should be converted into specific supply curve estimates, showing the amounts of future wood that could be produced under specified management and cost conditions, but I think that a detailed analysis would only confirm and quantify this general conclusion.

Potentials for Future National Forest Management

The National Forest Management Act of 1976 (Public Law 94-588) offers great potentials for future management of the national forests. It is likely to be recognized in the future as a historic major enabling act for the national forests. It is always difficult to estimate in advance exactly how any legislation will work out in practice. Much depends on how the agency concerned interprets the act, and upon the skill and energy the agency puts into the implementation of the act; but much also depends upon how the agency's Secretary, the Office of Management and Budget, the President, and the Congress control, restrict, or stimulate the agency, including the level of appropriations available for carrying out the act. This section of this report is therefore unavoidably somewhat speculative in character; nevertheless, there seems much advantage in trying to estimate in advance just how the National Forest Management Act of 1976 will operate in practice. To the extent an advance estimate of its possibilities is realistic, a later review or audit of the act's operations can measure how far the potentialities were achieved.

The 1976 act is formally an amendment to the Forest and Rangeland Renewable Resources Planning Act of 1974 (88 Stat. 476). The latter arose out of the concern by key members of the Congress and others that the full potentials of the renewable resources of the United States, and more particularly those of the national forests, were not being realized. As its title suggests, the 1974 act was primarily concerned with planning. An assessment of the state or condition of all renewable resources of forest and range, whatever their ownership, was provided for in the 1974 act; and a program for the national forests was required by it. Revision of the 1974 act was given urgency by a court decision in August 1975, known as the Monongahela case, which required the Forest Service to comply strictly with the language in the 1897 enabling legislation about sale of only "dead, physiologically mature, and large growth" trees, each of which had to be individually marked as available for cutting. That decision, if applied to all national forests, would have drastically reduced timber harvest from them. There was also a desire to strengthen some sections of the 1974 act, as well as to add some supplementary sections.

The 1976 National Forest Management Act, as indeed much national legislation, represents a compromise among many competing interests. The forest industry and the home-building industry were much concerned to maintain or to increase the flow of timber from the national forests, and they wanted the 1976 act to cure the Monongahela decision. Many conservationists were concerned to limit or control timber harvest on national forests for conservation reasons. Others were much concerned with the economics of national forest management and with the opportunity for public participation in the planning and operations of the national forests. These and other interest groups each naturally sought an act which would meet their objectives or one which at least was not harmful to their interests. Since the passage of the act in late October 1976, several of the divergent interests have suggested that the act is tolerable or better from their viewpoint. If that apparent consensus holds up under its actual operation, the act may well be a great success in terms of a reconciliation of divergent views.

The 1976 National Forest Management Act is long and complicated, and many sections or words or phrases are easily subject to different interpretations by different persons or groups. One may be sure that some sections of the act will be tested in court, as some group feels its interests are not properly served by the administration of the act. This may take several years. It is noteworthy that the 1897 act had been interpreted in one way by the Forest Service and by most if not all users of the national forests until the Monongahela case arose in the early 1970s. Pending such court cases, anyone's interpretation of the act is only tentative; and, as the Monongahela case well illustrates, court interpretations may well lead to new legislation to clarify what many persons had believed an act said.

The 1976 National Forest Management Act requires the Secretary of Agriculture (which in practice means the Forest Service) to take certain actions, prohibits or regulates him in the taking of other kinds of actions, and implicitly allows him much leeway on still other matters. Although the first two of these categories are not without the possibility of differing interpretations, the greatest likelihood of differences in interpretation lie in those things where some measure of discretion is authorized or implied. The analysis which follows is by kinds of activities but the degree to which each is mandated, forbidden, or left for discretion will be suggested. The following analysis of the 1976 act does not include all features of the act but only those parts more or less directly related to the subject matter of this report, namely the decision-making process with regard to timber harvest.

The 1976 act, as the 1974 act before it, strongly emphasizes planning, primarily for the national forests but to some extent for all renewable natural resources. Section 2, "Findings" emphasizes the national interest in preparing resource assessments and programs "...based on a comprehensive assessment of present and anticipated uses, demand for, and supply of renewable resources from the nation's public and private forests and rangelands..." Section 6 of the act spells out in some detail how public participation in the national forest planning process shall be provided for. "...plans for units of the national forest system shall... provide for multiple use and sustained yield of the products and services obtained therefrom in accordance with the Multiple-Use Sustained-Yield Act of 1960, and, in particular, include coordination of outdoor recreation, range, timber, watershed, wildlife, and fish, and wilderness". "Plans developed...shall (1) form an integrated plan for each unit of the national forest system,...(2) be embodied in appropriate written material,...(3) be prepared by an interdisciplinary team,...(4) be amended...and (5) be revised...".

The intent of these quoted sections, and to some degree of other sections not quoted here, is clearly to provide a comprehensive planning process for the national forests. The emphasis upon all outputs of the forests, upon integration among the elements of the plan, and for an interdisciplinary approach are noteworthy. Each of these terms is subject to more than one definition in its application, hence some differences may arise among persons concerned with national forests, as to how far the act has been implemented in this respect. The act refers to "units" of the national forest system; are these regions, forests or ranger districts, or some other unit? Again, differences in definition are possible.

These sections clearly require the Secretary of Agriculture (hence the Forest Service) to do this planning, while leaving much to their discretion as to how the planning should be carried out.

Other sections of the act relate to reporting. For instance, in section 3(c) the Secretary (Forest Service) is required in the Assessments provided elsewhere in the act to report "...the additional fiber potential in the National Forest System including, but not restricted to, forest mortality, growth, salvage potential, potential for increased forest product sales, economic constraints, alternate markets, contract considerations, and other multiple use considerations;...." While a section such as this does not explicitly require suggestions on how to realize the unrealized potential, it clearly points in that direction. Again in section 4(d)(1) the Secretary (Forest Service) is directed to report on "the amount and location by forests and states and by productivity class, where practicable, of all lands in the National Forest System where objectives of land management plans indicate the need to reforest areas that have been cut-over or otherwise denuded or deforested, and all lands with stands of trees that are not growing at their best potential rate of growth". Other parts of this same section require reporting of the success of reforestation activities and annual submission of estimates of the costs of remaining necessary reforestation.

The intent of these sections to have the national forests productively growing timber is clear. The reporting in itself does not grow trees, of course, but it sets the stage for their being planted.

The 1976 act requires the Secretary (Forest Service) to undertake a number of actions on the national forests. For instance, section 6(g) in its several subsections requires the Secretary within two years (or sooner if practicable) to "promulgate regulations, under the principles of the Multiple-Use Sustained-Yield Act of 1960, that set out the process for the development and revision of the land management plans, and the guidelines and standards prescribed by this subsection". The guidelines must "require the identification of the suitability of lands for resource management; provide for obtaining inventory data on the various renewable resources...and provide for methods to identify special conditions or situations involving hazards to the various resources and their relationship to alternative activities." The guidelines for land management plans shall "insure consideration of the economic and environmental aspects of the various systems of renewable resource management...; provide for diversity of plant and animal communities based on the suitability and capability of the specific land area...." The intent of these sections, it seems clear, is to promote the formulation of a carefully planned and well-integrated plan of management of the national forests, and to see that the plan is in fact carried out. The quoted language, and other parts of the act relating thereto, clearly leave great discretion in the hands of the Secretary as to how the mandated goals are to be achieved.

Section 6(1) specifies that "The Secretary shall (1) formulate and implement, as soon as practicable, a process for estimating longterm costs and benefits to support the program evaluation requirements of this act. This process shall include requirements to provide information on a representative sample basis of estimated expenditures associated with the reforestation, timber stand improvement, and sale of timber from the National Forest System, and shall provide a comparison of these expenditures to the return to the government resulting from the sale of timber." This is clearly a move in the direction of comparing costs and returns, although it does not include all costs of timber production (notably not interest on the value of the stand nor rent from the land) nor does it include values other than timber which arise from national forest management.

Section 6(m) requires the Secretary to "establish (1) standards to insure that, prior to harvest, stands of trees throughout the National Forest System shall generally have reached the culmination of mean annual increment of growth (calculated on the basis of cubic measurement or other methods of calculation at the discretion of the Secretary): _Provided_ That these standards shall not preclude the use of sound silvicultural practices, such as thinning or other stand improvement measures: _Provided further_, That these standards shall not preclude the Secretary from salvage or sanitation harvesting of timber stands which are substantially damaged by fire, windthrow or other catastrophe, or which are in imminent danger from insect or disease attack; and (2) exceptions to these standards for the harvest of particular species of trees in management units after consideration has been given to the multiple uses of the forest including, but not limited to, recreation, wildlife habitat, and range and after completion of public participation processes utilizing the procedures of subsection (d) of this section."

This subsection appears at first reading to be a complete requirement of maximizing MAI which earlier sections of this report have shown is generally uneconomic. The subsection does provide a number of escape hatches from the straight-jacket of maximum MAI--the exclusion of thinnings and salvage, and the possibilities of other management when necessary for other purposes. But most economists would wish this subsection was not in the act. However, other sections, as will be noted later, not only permit but encourage intensive forestry under appropriate conditions; and, as noted earlier in this report, the age-growth-yield curve will not only shift under intensive management but will likely become steeper in its rising phase, so that measurement of maximum MAI will become more difficult while at the same time the economic loss from achieving it will be less. One may well anticipate some lively arguments over whether maximum MAI has, or has not, been reached on a particular forest tract at a particular date.

Section 14 sets forth in considerable detail how the Secretary shall make timber sales from national forests. These include how timber sales shall be advertized and the terms of timber sale contracts. He is enjoined to "obviate collusive practices in bidding for trees,..." and specific requirements to this end are established. Planning and supervision of timber sales shall be entirely by persons employed by the Secretary. He is authorized and directed to establish utilization standards for the harvest of timber. In numerous other ways, the details of timber sales are regulated by this section of the act. While these regulations are detailed, in general they correspond fairly well with present practice in the Forest Service or with what present practice should be, and they should impose no unreasonable hardship upon the Forest Service.

The act also limits the Secretary (Forest Service) in a number of specific ways. For instance, Section 6(g)(3)(F) requires the Secretary to "insure that clearcutting, seed tree cutting, shelterwood cutting, and other cuts designed to regenerate an even-aged stand of timber will be used as a cutting method on the National Forest System only where--(i) for clearcutting, it is determined to be the optimum method, and for other such cuts it is determined to be appropriate, to meet the objectives and requirements of the relevant land management plan; (ii) the inter-disciplinary review as determined by the Secretary has been completed and the potential environmental, biological, esthetic, engineering, and economic impacts on each advertised sale area have been assessed, as well as the consistency of the sale with the multiple use of the general area; (iii) cut blocks, patches, or strips are shaped and blended to the extent practicable with the natural terrain; (iv) there are established according to geographic areas, forest types, or other suitable classifications the maximum size limits for areas to be cut in one harvest operation...; and (v) such cuts are carried out in a manner consistent with the protection of soil, watershed, fish, wildlife, recreation, and esthetic resources, and the regeneration of the timber resource." These provisions of the act are basically the so-called Church guidelines, named after Senator Frank Church, of Idaho, when a Senate subcommittee under his chairmanship promulgated these rules some years ago. The Forest Service accepted the Church guidelines and has been following them for some years. Their specification in law does, however, give them added importance and added legal position, a circumstance which is pleasing to many of the conservation groups. At the same time, clear-cutting as a method timber harvest is now fully authorized, a circumstance pleasing to the Forest Service and to purchasers of national forest timber.

The act also limits the Secretary (Forest Service) in making timber sales from certain kinds of national forest areas. Section 6(k) states "In developing land management plans pursuant to this Act, the Secretary shall identify lands within the management area which are not suited for timber production, considering physical, economic, and other pertinent factors to the extent feasible, as determined by the Secretary, and shall assure that, except for salvage sales or sales necessitated to protect other multiple use values, no timber harvesting shall occur on such lands for a period of 10 years....The Secretary shall review his decision to classify these lands as not suited for timber production every 10 years and shall return these lands to timber production whenever he determines that conditions have changed so that they become suitable for timber production". This limitation is similar to the suggestion earlier in this report that most site class V and some site class IV "commercial" forests in the national forests are not truly commercial in their economic potential and thus should not be harvested, at least at present. This provision of the act is broad in the sense that many different kinds of considerations are listed for the Secretary's decision. Clearly there may well be differences of opinion in the application of these criteria to specific national forest situations, and it will be most interesting to see how the Forest Service applies this section of the act.

Section 11 of the act also limits the Secretary (Forest Service) in the amount of timber that may be sold annually from the national forests. "The Secretary of Agriculture shall limit the sale of timber from each national forest to a quantity equal to or less than a quantity which can be removed from such forest annually in perpetuity on a sustained yield basis: Provided, That, in order to meet overall multiple-use objectives, the Secretary may establish an allowable sale quantity for any decade which departs from the projected long-term average sale quantity that would otherwise be established....In addition, within any decade, the Secretary may establish a quantity in excess of the annual allowable sale quantity established pursuant to this section in the case of any national forest so long as the average sale quantities of timber from such national forest over the decade covered by the plan do not exceed such quantity limitation....Nothing in...this section shall prohibit the Secretary from salvage or sanitation harvesting of timber stands which are substantially damaged by fire, windthrow, or other catastrophe, or which are in imminent danger from insect or disease attack." This section of the act goes further in establishing even flow (without using this phrase) as a restrictive variant of sustained yield than any economist would consider economically justified, and to this extent seems regrettable. The quoted language does provide some latitudes or escape hatches to prevent undue rigidity in year to year operations and to allow for salvage and other sound timber management practices. More important than the exceptions spelled out in this section is the fact that other sections of the act seem rather clearly to encourage, or at least to make possible, intensive forestry where this is ecologically and economically sound. Intensive forestry can make possible a much higher annual turnoff, even under the terms of this section, that is possible with "natural" forestry, as our earlier examples have shown. The practical effect of section 11 will thus depend upon how the Forest Service proposes to manage timber growing and harvest, and upon how well the Secretary, OMB, the President, and the Congress support the Forest Service plans.

As I read the act, I am impressed with the opportunities it offers the Secretary (Forest Service) not so much explicitly as implicitly, for methods of timber management on the national forests that I would consider desirable. I see considerable incentive in the act for intensive forestry on the national forests where this will be ecologically and economically sound. By such measures, the estimated sustained yield can be increased considerably. The Secretary is not required to harvest timber everywhere on the national forests where trees may grow; on the contrary, he is explicitly forbidden to harvest under some conditions, as noted above, and consideration of other outputs from the forests under programs of multiple use management may well preclude timber harvest in other situations. He is constrained to harvest at maximum MAI and to harvest on an even flow (subject to the exceptions noted above), and these are constraints I would have preferred to have omitted from the Act. But the effect of these constraints can be significantly minimized by intensive forestry, as we have noted.

One feature of the act which has intriguing possibilities is the authorization (section 14(a)) for the sale of logs as well as the sale of stumpage from the national forests. If used, this authority would make possible the sale of logs at log markets, which has important conservation as well as economic potential. In particular, the thinning of forest stands and the sale of the logs so removed would be a great deal easier if logs rather than standing trees were sold.

In summary, only some of the more important features of the act, particularly those relating to timber growing and harvest, have been discussed here. As pointed out at the beginning of this section, the act is long and complicated, and not everyone will agree with any interpretation of it. More importantly, a final judgment cannot be reached until the act has been in operation for some years. But the act is clearly important both for what it says and for what it implies, and it may well mean a new day in national forest management. Even a preliminary judgment must take into account the tone and thrust of the whole act, as well as the wording of separate portions of it. When so considered, I am much impressed with the vistas that the act opens up for the future.

Conclusion

This report has dealt with the decision-making process in timber production, harvest, and marketing. The important relations between timber and the other outputs of the forest are considered only incidentally since I have dealt with them more fully elsewhere. The decision-making process is considered for three major types of forest owners/managers: small private forests, large integrated forest-owning and processing firms, and the national forests. For each, the decision-making process is considered spearately for the harvest of existing stands of timber and for the growing of more timber, and for the national forests the discussion is in terms of past and present management, and of future management possible under the National Forest Management Act of 1976.

All of the decisions are based, explicitly or implicitly, upon the time-growth-volume biological relationships shown in table 1 and in figures 1 and 2. Although only one such relationship is shown in the table and figures, in fact there is a family of such relationships for each site, as management differs and as inputs other than the site vary.

There is one special point--or, given the unavoidable lack of precision in measurement, one approximate area--on each such curve, where the MAI of growth is at a maximum. Although this is but one of several points that might be defined on such curves, this point is widely used by foresters. This point does have the special advantage that harvest at that age will maximize the total physical volume of wood growth over several rotations on the same site. But this point has meaning only for a defined method of management of timber on the site. A different management program may well produce maximum MAI at a different age; in particular, intensive forestry is likely to produce a higher maximim MAI, at an earlier age, and to result in a longer span of the curve where maximum MAI is hard to determine and may in fact be indeterminate.

All decisions about timber production, harvesting, and marketing include economic as well as biological considerations, at least implicitly and at least to some degree. Even harvest at maximum MAI is based upon estimates of harvestable volume of wood on the site, and this in turn is based upon the economics of utilizing wood of different dimensions. There is one special point (or, as with MAI, more probably an area) in the economic relationships of time-growth-volume which is of special interest to economists--the point of financial maturity, where the wood grown to that age has the greatest value in relation to the costs of growing it to that age. The matter of costs is not simple; in particular, how to include interest on the capital in the timber stand and how to include rent on the site typically present difficult problems. In practice today financial maturity is reached at a considerably younger age than is maximum MAI, and a considerably lower cost per unit of wood grown.

A major conclusion of this study has been that the biological and economic relationships based on a time-growth-volume curve (whether maximum MAI, financial maturity, or any other) are not sufficient for making decisions about timber harvest. Neither the biological nor the economic points are precise; instead, there exists a period of years, sometimes a relatively long one, within which the probable exact year of maximum MAI or of financial maturity falls, and during which harvest may take place at any year with approximately the same results. But beyond this imprecision lie other factors which should be, but typically are not, included in either a biological or an economic analysis of optimum harvest age. For the small private forest owner, the state of the market will generally be decisive in decisions about when to harvest an existing stand of timber. His objectives in forest ownership will be decisive in decisions about further investment to grow more timber than will grow naturally with a minimum of investment. For the large integrated forest industry firm and for the national forests, the effect of harvest decisions on market price must be considered; while this can be done under some advanced economic models, it often is not considered. For the private firm, the economics of timber processing may be as important, or more so, than the economics of wood growing. The Forest Service has traditionally paid great attention to community stability and its judgment of timber harvest to meet this condition may override economic and biological analysis.

The last major section of this report suggests that in the future the management of the national forests will be based more directly on biological and economic considerations than it has been in the past.

The National Forest Management Act of 1976 requires that timber reach maximum MAI before it can be harvested, which is an obvious reliance on the biological yield curve (with all the difficulties this entails in practice). The act also requires a particular form of sustained yield harvest of timber, thus a reliance on economics as well as on biology, even though most economists would reject this particular form of economic test. But the act is a major advance over the past, in its requirements for an explicit analysis of management plans and the strong emphasis on management planning. The act both gives the Secretary (Forest Service) clearer guidelines than does any earlier legislation and at the same time allows the Executive Branch great discretion in reaching the goals and aims of the act. More important than the several explicit guidelines is the implicit encouragement, backed up where necessary with explicit provisions, for improved management, higher output, more intensive timber management, better matching of costs and returns, and generally higher output at more explicitly economic costs. The act offers great possibilities which it is hoped will be realized in practice.

APPENDIX. DETERMINATION OF THE ECONOMICALLY OPTIMUM AGE OF TIMBER HARVEST

Pages 22 to 32 of this report have presented a general nontechnical discussion of the problem of determining the economically optimum age of harvest for a timber stand. Over a period of more than one hundred years there has been developed an enormous volume of technical literature on this subject. Merely to list all of the various reports would be a very large job, and one would almost surely miss some. A full analysis of all these reports is probably impossible for any single author, and even if possible would be wholly out of keeping with the purpose and style of this report. My goal in this Appendix is much more modest. The pages which follow present only the barest minimum outline of some of the recent economic literature on this subject. They are designed primarily to give the interested reader a little flavor of the literature and some references by which he or she can follow the subject further, if interested. This discussion may perhaps be most useful to the beginning student in forestry or in environmental management or to one who wishes to know something about the subject without getting too involved.

As a useful generalization, the literature can be divided very broadly into two categories: (1) that by professional economists interested primarily in economic theory, especially in capital theory, who are generally much less well informed about forestry than about economics, and who are generally less interested in the application of their theory than in its development; and (2) that by forest economists who are generally competent foresters as well as economists, interested in the application of theory to practical forestry problems more than in the development of theory for its own sake. There is clearly no sharp line dividing these two groups; some observers might classify a writer into a different group than would some other observer; and some writers almost defy classification. Nevertheless, this classification is useful and is employed in this Appendix.

It is useful to start a review of recent economic literature, of the first of the foregoing two broad classes, by reference to Gaffney's work [4]. He neatly summarizes some of the major concepts of timber harvest determination into a single table (table 3). This table and his report analyze the writings of R. G. D. Allen, Irving Fisher, Kenneth Boulding, and others, all better known as economists than as foresters. Basically, Gaffney revives the findings of Martin Faustmann, who wrote in 1849 and whose work is often the starting point for all writers on this subject. Gaffney expands and generalizes Faustmann's approach.

In the derivation of his formulas, Gaffney initially makes the assumption that all costs of timber growing are incurred at the beginning of a rotation--"regeneration"--in his terms; and that all income is received when the trees are cut. Later, he shows how recurrent costs and intermediate income can be worked into his basic formulas, but he does not make much use of such intermediate costs and returns. Gaffney develops several formulas, under varying assumptions, which are too detailed to be reproduced here. His expression of the Faustmann formula (in Gaffney's terminology) is as follows (formula 6, pp. 7 and 9):

Table 3. Some Concepts of Financial Maturity and Their Advocates in Relation to Two Elements of Urgency

If the Site Rent is → / If the Interest Rate is Equal to ↓	Zero	Annual Equivalent of Periodic Net Revenue
	Financial Maturity is a Maximum of:	
Zero	Total Growth is Maximum (No avowed Advocates)	Mean Annual Net Growth (Waldrente)
Market Rate	Discounted Net Revenue is Maximum (Allen and Fisher)1/	Annualized Net Revenue (Bodenrente) (Faustmann)3/
Maximum Internal Rate	"Internal Rate of Return is maximum (Boulding)2/	Site Rent and Internal Rate of Return (present study)4/

Adapted from Chart, p. viii, Gaffney, see [4].

1/ Allen, R.G.D., Mathematical Analysis for Economists, London: MacMillan and Co., 1938. Fisher, Irving, The Theory of Interest, New York, MacMillan Co., 1930.

2/ Boulding, Kenneth, Economic Analysis, New York, Harper and Bros., Third Edition, 1955.

3/ Faustmann, Martin, Berechnung des Werthes, welchen Waldboden sowie noch nicht haubare Holzbestande fur die Waldwirtshhaft besitzen", Allg. Forst Und Jagd-Zeitung 25: 441-455, 1849. Translated into English as Martin Faustmann and the Evolution of Discounted Cash Flow, by M. Gane, Commonwealth Forestry Institute Paper No. 42, University of Oxford, 1968.

4/ Gaffney, M. Mason, Concepts of Financial Maturity of Timber and Other Assets, A.E. Information Series No. 62, Department of Agricultural Economics, North Carolina State College, Raleigh, North Carolina, December 1960.

$$t = \frac{1}{p} \ln \frac{g'}{g' - p(g - C_o)}$$

where t is the length of the timber rotation (measured in years);

p is the interest rate, measured continuously throughout the year, which compounded equals the annual interest rate (a refinement which even Gaffney admits usually makes no great difference, as contrasted with use of an interest rate);

g' is the annual increment in stumpage value;

g is the total stumpage value; and

C_o is the regeneration cost.

Gaffney discusses numerous approximations of shortcuts and some formulas advanced by other authors (including some not listed in table 3) and shows how each of these may be reasonably good, and perhaps simpler than his own approach, but always under some special circumstances but not under others. For instance, the land rent may be zero or close to it, or the supply of capital may be unlimited at a known interest rate so that the returns to some factor other than capital should be maximized, or other special circumstances may prevail.

Gaffney compares the results of his (or the Faustmann improved) formula with the results obtained by others, in a number of particular situations. Sometimes, the results are nearly the same, sometimes they are quite different, and where results differ, he naturally feels that his analysis is the more accurate. As noted, some approximations do very well in some situations, but discrepancies may be evident only when the correct formula is used as a basis of comparison. As he notes (p. 89): "But Faustmann's formula is not final solution to the question of financial maturity. It is only a simple classic theme, if you will, for latter day composers to weave into larger patterns, to vary and orchestrate with the full symphony of modern instruments".

A recent paper by Paul Samuelson [25] deserves special mention, partly because of its recentness and partly because of the professional stature of its author. Samuelson reviews a number of studies, including that by Faustmann and several by some of the same authors whose work Gaffney had analyzed more than fifteen years earlier, some of which are listed as sources for table 3. All or nearly all the writers Samuelson refers to are better known as economists than as foresters. While Samuelson mentions Gaffney, he does not put much stress on his analysis, preferring to go back to the original works which Gaffney had analyzed. Samuelson goes on to develop his own solution to the problems of optimum age at which to harvest timber, in a highly elegant and fully rigorous way that bears some resemblance to Gaffney and Faustmann. He is very careful to state the four "definite assumptions" upon which his analysis rests, as follows:

"(1) knowledge of future lumber prices at which all outputs can be freely sold, and future wages of all inputs;

(2) knowledge of future interest rates at which the enterprise can both borrow and lend in indefinite amounts; and

(3) knowledge of technical lumber yields that emerge at future dates once certain expenditure inputs are made (plantings, sprayings, thinnings, fellings, etc.).

Finally, it is assumed (4) that each kind of land suitable for forests can be bought and sold and rented in arm's length transactions between numerous competitors; or, if the government owns public lands, it rents them out at auction to the highest of numerous alternative bidders and conducts any of its own forestry operations so as to earn the same maximum rent obtainable at the postulated market rate of interest." (Underlining in the original.)

To the economic analyst working on forestry problems, as to the forest owner/manager, these conditions are highly limiting. Samuelson recognizes that his assumptions are "heroic" but he defines them as necessary for a "simple" solution. On the basis of these assumptions, and by a rigorous mathematical deductive process, Samuelson comes up with a formula for deciding upon the proper timing of timber harvest, which includes input costs (which he reduces to labor costs), interest rate, and rent on land. "The first line of (6b) is the correct Faustmann-Gaffney-Hershleifer formulation. Its equivalence with the maximum-land-rent formulation of (6a) is seen from solving the last relation of (6a) for R and noting its equivalence with the second relation of (6b) except for the extraneous constant r."

$$\text{(6b)} \quad \underset{T}{\text{Max}}\, [Pf(T)e^{-rt} - WL]\, [1+e^{-rt} + (e^{-rt})^2 + \ldots]$$

$$= \underset{T}{\text{Max}}\, [Pf(T)e^{-rt} - WL]/[1-e^{-rt}] = \underset{T}{\text{Max}}\, (R/r)$$

$$= [Pf(T_\infty)e^{-rT_\infty} - WL]/[1-e^{-rT_\infty}] = R_\infty/r,$$

In this formula:

T = time

"In the steady state, a new part of each forest is being planted at every instant of time, an old part is being cut down at age T, and forest stands of all ages below T are represented in equal degree".

P = prices for "lumber," by which here, as elsewhere, he apparently means all wood products.

$f(T) = Q_t + T$ = quantity of "lumber" that can be cut at time T + increment of time t

"...as we'll see, f(T) is short for f(s,L,T), where L is labor input at the beginning of one planting cycle and s is the land used throughout that complete cycle".

r = interest rate

W = wages

L = quantity of labor input

} and by these Samuelson apparently includes all current inputs into the forest productive process

Samuelson agrees that if the land rent is zero, then this term can be dropped from his formula. He does not say, but presumably would agree, that if the input costs to the forest landowner were zero, then this term could also be dropped. If an owner did not seed or plant his forest land after harvest, but relied instead upon natural reseeding; if he did not fertilize, thin, or spray at any time; and if fire protection were entirely a public responsibility, then his input costs would indeed be zero or very close to it. Something not too unlike this exists for some small private forest owners. If input costs and rent are dropped out, then the cost associated with timber age are reduced to interest on the harvestable value of the stand. One can either use an assumed interest rate, and calculate an age of trees which maximizes returns above this interest rate, or one can calculate the age of trees at harvest which maximizes his rate of return. The results, under this particular set of conditions, should be the same.

Charles Howe, in a recent paper in the same general tradition, has also dealt with this same problem, in a somewhat simpler way, but stating that he is basing his analysis on those of Faustmann, Gaffney, and Samuelson [26]. He starts (p. vi) with a flat conclusion: "In forestry, as in fisheries, there are many levels of sustained yield. The question really concerns the socially optimal rotation, for once the rotation (or pattern of rotations) is set, the yields also are set". He goes on (p. 14): "For expositional purposes, let us assume that we are dealing with a fully regulated forest with an established rotation of t units of time (e.g., decades). A fraction of the forest area (1/t) might be cut and replanted during each unit of time. Letting f(t) represent the volume of timber standing on an acre of land as a function of the time since planting, the physical yield each unit of time would be:

$$(1) \qquad Y(t) = \frac{N}{t} \cdot f(t)$$

where N is the number of acres in the forest. This represents the simplest notion of sustained yield, a yield invariant over time (there is no implication that a forest thus "regulated" is in any sense optimal). (1) makes it clear that there exists a different 'sustained yield' for each value of t, i.e., for each possible rotation".

Howe continued with a more complex case:

"For one acre of forest land which has just been cut and replanted, the optimal rotation problem can be stated as

$$(2) \qquad \max_{t} \left\{ (\pi_o(t) = [p \,.\, f(t)e^{-rt} - k] \, [1 + e^{-rt} + e^{-2rt} + \ldots] \right\}$$

where π represents the present value of one acre, p the unit value of timber net of harvesting and transport costs, k for the replanting cost, and where it is assumed that the rotation will be constant forever". One can welcome his recognition that trees standing in a forest are not logs, much less that they are not lumber, and that the woods are not the point of consumption of the wood grown. There is no recognition of costs other than replanting, nor of management practices other than selecting the length of the rotation.

When it comes to the forest economists who were more interested in solving practical forest management problems than in developing a body of theory, there is also a large body of literature. In general, these writers have not spelled out their theory in as much detail as the group I call the economic theorists nor have they relied as much on deductive reasoning, but rather they have focused their energies on the application of theory. While these writers vary greatly, of course, most of them use economic theory which is fully adequate for the problem they address, but they do not consider theory as the end product of their work.

When it comes to this group of writers, one scarcely knows where to begin. Duerr in his textbook on forest economics, especially in chapter 9 utilizes this approach to practical forest management problems [27]. In doing so, he built on earlier research that he and colleagues had conducted [28]. In a new textbook published in 1975, Teeguarden has a chapter, number 17, largely devoted to this subject [29]. Each of these sources lists a number of references, to which the interested reader is referred. While this by no means includes all the literature up to the date of the respective publication, it does include some of the better known studies.

John Walker in a relatively recent study has made two major improvements over most earlier formulations: his model explicitly considers markets and the effect of harvest in one period of time upon timber prices in the next period of time, and his model "can start with a timber inventory that is non-optimal. This situation is commonly encountered where forest management must begin with an overmature stand of timber". [6]. "The relationship that must exist between the last unit of wood cut at time t, and the first unit cut at time t + 1, is expressed by the following first degree difference equation:

$$V_t(MR_t - MC_t)(1+i) = V_{t+1}(MR_{t+1} - MC_{t+1})$$

where V_t = the volume per acre of the last (youngest) age class harvested at time t,

V_{t+1} = the volume per acre of the forest (oldest) age class harvested at time t+1,

MR_t = the marginal revenue of the last unit of wood harvested at time t,

MR_{t+1} = the change in total revenue at time t+1 if a unit of wood of the first (oldest) age class is not harvested in t+1,

MC_t = the marginal cost of the last unit of wood harvested at time t,

MC_{t+1} = the change in total cost at time t+1 if a unit of wood of the first (oldest) age class is not harvested in t+1,

i = the rate of interest for equating net cash flows at times t and t+1.

K. Norman Johnson has reviewed the application of Walker's methods, by Walker and others, and by himself and associates, as applied to some illustrative national forest situations [30]. Johnson's analysis is more detailed and technical than is appropriate to include here, and is not easily summarized. He does point to some problems in the application of Walker's methodology, although he clearly feels that the objective of developing a method which can both use market effects and can start from the non-optimal forest situation is a valuable one. Johnson's analyses and those which he cites have major operational significance for the forests, both public and private, to which they apply.

William F. Hyde has developed the traditional models of economic analysis of optimum date of timber harvest a great deal further than any earlier analyst, primarily by the introduction of a number of additional factors which are highly important in actual forest management situations [7]. Specifically, his analysis is either innovative or stronger than earlier ones in the following ways: (1) by its explicit use of timber yields by site classes; (2) by showing the effects of various forest management practices, including thinning and fertilization, on timber yields, both as measured by volume of wood produced and as measured by size of trees at various stages of the growing cycle; (3) by its thorough analysis of the effects of lags in stand regeneration, on both timber yields and costs of timber production; (4) by a great deal of new material on costs of timber production by site classes, including the risks of failure in planting seedlings, cultural measures such as weeding, and others; and (5) its explicit introduction of differences in "access" or costs of roads and of transport, for sites in different locations and of different topography.

Hyde's analysis is not easily summarized in this brief Appendix. Perhaps more than any other author listed, his work bridges the theoretical and the practical approaches. He develops numerous formulas in the best theoretical tradition, but he also introduces a great deal of quantitative empirical data, and he draws policy conclusions of major importance. His analysis not only solves the standard problem of economic age of timber harvest under a wide range of natural and economic conditions, but it also provides the basis for a choice of the economically optimum intensity of timber management under specified conditions, and it also provides the basis for rational economic choice in the allocation of timber land among its many possible uses (such as timber harvest versus recreation).

Forest economists in countries other than the United States have studied the same problems as discussed in this report. A full review of literature from outside of the United States is impossible for this report. One easily read and directly applicable study is by Hiley[31]. His analysis is direct, practical, and realistic. While he gives primary attention to forest plantations, including their establishment, thinning, and other management matters, his analysis is very much in the tradition of forest economics as applied to all forest management problems.

REFERENCES

1. Clawson, Marion, Forests for Whom and For What?, Johns Hopkins University Press for Resources for the Future, Baltimore, 1975.

2. Clawson, Marion, editor, Forestry Policy for the Future: Conflict, Compromise, Consensus, Resources for the Future, Washington, 1974.

3. Hirshleifer, J., "'Sustained Yield' versus Capital Theory", an unpublished paper presented at a symposium, "The Economics of Sustained Yield Forestry", University of Washington, Seattle, November 1974.

4. Gaffney, Mason, "Concepts of Financial Maturity of Timber and Other Assets", Agricultural Economics Information Series 62, North Carolina State College, Raleigh, N.C., September 1957.

5. Marty, Robert J., "Economic Effectiveness of Silvicultural Investments for Softwood Timber Production", Appendix D, Report of the President's Advisory Panel on Timber and the Environment, GPO, Washington, April 1973.

6. Walker, John, "An Economic Model for Optimizing the Rate of Timber Harvesting", Ph.D. thesis, University of Washington, Seattle, 1971.

7. Hyde, William F., "Timber Supply and Forest Land Allocation" unpublished manuscript, Resources for the Future, Washington 1977.

8. Data provided by Don Lee Fraser, of the Washington Department of Natural Resources.

9. U.S. Forest Service, The Outlook for Timber in the United States, GPO, Washington, 1973.

10. Mead, Walter J., Competition and Oligopsony in the Douglas Fir Lumber Industry, University of California Press, Berkeley and Los Angeles, 1966.

11. Report of the President's Advisory Panel on Timber and the Environment, GPO, Washington, 1973, p. 81.

12. Dana, Samuel T., Forest and Range Policy: Its Development in United States, McGraw-Hill Book Co., New York, 1956, pp. 391-396.

13. Gates, Paul W., History of Public Land Law Development, GOP, Washington 1968, p. 580.

14. Innovations and Trees - Weyerhaeuser; 1900-1975, written and published by Weyerhaeuser Company, Tacoma, Washington, 1975.

15. Clawson, Marion, "The National Forests", Science, Volume 191, Feb. 20, 1976.

16. Clawson, Marion, The Federal Lands Since 1956: Recent Trends in Use and Management, Resources for the Future, Washington, 1967, p. 28.

17. Robinson, Glen O., The Forest Service--A Study in Public Land Management, Johns Hopkins University Press for Resources for the Future, 1975.

18. Forest Service Manual 2410-2419 (May 1972).

19. Clawson, Marion, The Economics of National Forest Management, Resources for the Future, Washington, June 1976, See pp. 19-21.

20. Clawson, Marion and Held, Burnell, The Federal Lands: Their Use and Management, Johns Hopkins University Press for Resources for the Future, 1957, see pp. 62-66.

21. For an explanation and defense of this idea, and some references as to its history, see Dennis L. Schweitzer, Robert W. Sassaman, and Con H. Schallau, "Allowable Cut Effect--Some Physical and Economic Implications", Journal of Forestry, Volume 70, Number 7, July 1972, pp. 415-418.

22. Biological Productivity of Renewable Resources Used as Industrial Materials, Panel on Productivity, Committee on Renewable Resources for Industrial Materials, Board of Agriculture and Renewable Resources, Commission on Natural Resources, The National Research Council--National Academy of Science, Washington, D.C. 1976.

23. Clawson, Marion, and Hyde, William, "Managing the Forests of the Coastal Pacific Northwest for Maximum Social Returns", Timber Policy Issues in British Columbia, Edited by William McKillop and Walter J. Mead, University of British Columbia Press, 1976; and op.cit., Clawson, Forest Policy.

24. Vaux, Henry J., "How Much Land Do We Need for Timber Growing", Journal of Forestry, July 1973.

25. Samuelson, Paul A., "Economics of Forestry in an Evolving Society", Economic Inquiry, Volume XIV, No. 4, December 1976.

26. Howe, Charles W., "Economic and Social Perspectives Relevant to Forest Policy", unpublished manuscript submitted to Pacific Northwest Forest and Range Experiment Station, Forest Service, U.S. Department of Agriculture, February 1976.

27. Duerr, William A., Fundamentals of Forestry Economics, McGraw-Hill Book Company, New York, 1960.

28. Duerr, William A., John Fedkiw, and Sam Guttenberg, Financial Maturity: A Guide to Profitable Timber Growing, USDA, Technical Bulletin 1146, 1956.

29. Duerr, William A., Dennis E. Teeguarden, Sam Guttenberg, and Neils B. Christiansen, Forest Resource Management, published by Oregon State University Book Stores, Inc., Corvallis, Oregon, 1975. See particularly Chapter 17.

30. Johnson, K. Norman, "Consequences of Economic Harvest Scheduling Procedures on Five National Forests", an unpublished study, scheduled to be published by Utah State Agricultural Experiment Station, Logan, Utah.

31. Hiley, W.E., Economics of Plantations, Faber and Faber Limited, London, 1956.